# BASIC Microcomputing and Biostatistics

# BASIC Microcomputing and Biostatistics

*How to Program and Use Your Microcomputer for Data Analysis in the Physical and Life Sciences, Including Medicine*

WITH A PERMANENT LIBRARY OF MANY HELPFUL PROGRAMS, CHALLENGING PROBLEMS, AND WORKED EXERCISES

by

**Donald W. Rogers**

**Humana Press Inc • Clifton, New Jersey**

To

**Doris M. Rogers**

With other men, perhaps, such things would not have been inducements; but as for me, I am tormented with an everlasting itch for things remote. I love to sail forbidden seas, and land on barbarous coasts.

**—Melville**

**Library of Congress Cataloging in Publication Data**

Rogers, Donald, 1932-
Basic microcomputing and biostatistics.

(Contemporary instrumentation and analysis)
Includes bibliographies and index.
1. Life sciences—Data processing. I. Title.
II. Series.
QH324.2.R63 574'.028'54 81-85465
ISBN 0-89603-015-6 AACR2


Crescent Manor
P.O. Box 2148
Clifton, NJ 07015



Printed in the United States of America.

# Preface

*BASIC Microcomputing and Biostatistics* is designed as the first practical "how to" guide to both computer programming in BASIC and the statistical data processing techniques needed to analyze experimental, clinical, and other numerical data. It provides a small vocabulary of essential computer statements and shows how they are used to solve problems in the biological, physical, and medical sciences. No mathematical background beyond algebra and an inkling of the principles of calculus is assumed. All more advanced mathematical techniques are developed from "scratch" before they are used. The computing language is BASIC, a high-level language that is easy to learn and widely available using time-sharing computer systems and personal microcomputers.

The strategy of the book is to present computer programming at the outset and to use it throughout. BASIC is developed in a way reminiscent of graded readers used in human languages; the first programs are so simple that they can be read almost without an introduction to the language. Each program thereafter contains new vocabulary and one or more concepts, explained in the text, not used in the previous ones. By gradual stages, the reader can progress from programs that do nothing more than count from one to ten to sophisticated programs for nonlinear curve fitting, matrix algebra, and multiple regression. There are 33 working programs and, except for the introductory ones, each performs a useful function in everyday data processing problems encountered by the experimentalist in many diverse fields.

*BASIC Microcomputing and Biostatistics* arose as the result of a twofold need perceived by the author almost a decade ago. Students of statistics were overwhelmed by masses of trivial calculations that were meant to illustrate statistical principles, but obscured them instead. Even routine techniques, such as constructing simple and cumulative frequency distributions, were made complicated by the need to sort many numbers into arbitrarily sized "bins" while wondering what to do with the data that fall on the border separating one bin from its neighbor. On the other hand, in computer programming courses, then rare, now ubiquitous, programming principles were taught using exercises that illustrated them more or less well, depending on the textbook, but solved no problem anyone needed or wanted to solve.

Statistics courses needed a means of handling routine number crunching so that the student might see the conceptual forest for the trees and neophyte computer programmers needed meaningful problems to solve with their newfound programming skills. Why not combine statistics and programming and teach two courses in one? I tried it; it worked. The course outline became the framework for this book and it has been filled in and modified from then until now.

Taught in its entirety, this book is intended as a text for a one-semester course in biostatistics, biomedical statistics, or related fields such as quantitative courses in agriculture, agronomy, ecology, botany, experimental psychology, indeed any field in which masses of numerical data are treated. Students have finished one semester of this combined course with an understanding of statistics roughly equivalent to the level attained in the traditional one-semester statistics course, and they have mastered the fundamentals of computer programming in BASIC at about the level achieved by students in the usual one-semester course in that subject. Not the fine points, mind you, but enough to make the machine work for them, solving the quantitative problems that life scientists encounter in the laboratory or in the field.

Although complete coverage of this book is at the advanced undergraduate–beginning graduate level, experience has shown that, by selective deletion, it can be modified for use in quite a broad spectrum of programming courses. By cutting out all the calculus (presenting the integral as merely an area and giving several equations *ex cathedra*) I have used the manuscript as a basis for a crash review of high-level programming for juniors who are about to embark on an assembly-language course, a one-week short course for junior biologists entering an NIH research program, the computer component of an honors course in introductory chemistry, and a similar honors course for beginning biologists. By deleting almost all of the theory and concentrating on games and graphics, I have had considerable success using programs in this book in a two-and-a-half week orientation course for entering freshmen at L.I.U. in the summer of 1982. All the programs in this book have been run on a DEC 20-60 time-sharing system. With a very few modifications, discussed in the text, they have also been run on a home microcomputer. These short, simple programs are ideally suited to bring the microcomputer into play as a powerful laboratory research tool.

# Table of Contents

# Chapter 1

# Introduction to the Computer

It is by no means necessary for the sometime computer programmer to have any understanding of what goes on inside the machine he or she controls. However, a brief sketch of computer architecture and operation will make some of the nomenclature to be used in future chapters easier to understand.

The analogy to driving an automobile illustrates the way most people use computers. Although luxuries and refinements have made the detailed workings of a present-day automobile hopelessly complicated for anyone but a professional mechanic, it is basically a very simple machine, working on mechanical principles known to the Greeks: the wheel and the lever. When we drive, we never give a thought to how these components are functioning, we simply give a series of commands to the machine. Commands are transmitted from the driver to the machine through *input devices*. The steering wheel transmits the driver's wish to change direction, the accelerator and brake transmit the desire to change speed, and so on.

The computer, although it may not seem so, is also a very simple machine. To run it, we must transmit our desires to the machine through one or more of various *input devices*: a card reader, a teletypewriter, a tape reader, or others. The sequence of commands to a computer—the *source program*—is different for each new job, just as the sequence of commands to an automobile is different for each new destination. In one respect, computer programming has an advantage over driving a car. Once a program has been written for solving a given problem by computer, it can be run dozens or hundreds of times for different sets of experimental data. Drivers, however, must change the set of commands they input to their machines each time they traverse a given route to account for changing traffic, weather, and other conditions. A successful source program can be run any number of times without change.

The instructions that constitute the source program are executed in the part of the computer called the central processing unit, almost universally abbreviated CPU. Once the computer has done the jobs its programmer has directed it to do, it has the answer to the problem in its *memory*. That does not do us much good, however, until we get it out in comprehensible form. This is accomplished by means of *output devices* such as the on-line printer, the terminal printer, which looks like a typewriter, or the cathode ray tube (CRT), which looks like a TV screen. Computer programs may be permanently stored for future use by such devices as a card punch, which stores the program in encoded form as a series of holes punched into a deck of the familiar 3-¼ by 7-¼ inch computer cards. Other storage mechanisms, such as disc and magnetic tape are widely used.

Devices for inputting and outputting information are called *I/O devices*, but since they are not central to the operation of the computer, they are also called *peripherals*. A simple block diagram shows the operation of a computer as we have described it so far.

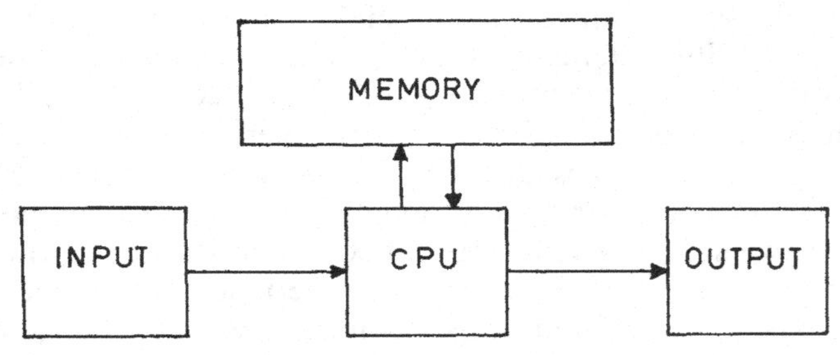


Fig. 1-1. Simple block diagram of a computer.

## Computer Memory

There are many mechanical and electronic ways to store information. The Chinese abacus and the mechanical adding machine are examples. Many computers use small ferromagnetic rings that can be magnetized in either one of two possible ways, up or down. Since there are only two stable states, these magnetic rings or *cores* are called *bistable devices*. There are also other bistable devices such as switches and relays used in computers. In recent years, memories based on transistors called monolithic *random access memories* or simply RAMs have become very popular, particularly in smaller machines.

Suppose we agree that one state of a bistable device symbolizes the number 0 and the other state symbolizes the number 1. Suppose further that we have the capability to reverse the state of our device by means of an electrical pulse and that we have a means of determining which state it is in, the *0 state* or the *1 state*. If we start out with the device in the zero condition and do not pulse it electrically, we retain the zero state stored in that device. If we pulse it, the field is reversed and the device is sent into the *one* condition. We now have a way of storing information. We have either a 0 or 1 stored in the device and any time we come back with our detector, we shall find the number we stored. The piece of information stored in one device is called a *bit* (contraction of *bi*nary dig*it*).

Not very spectacular, you say? Perhaps not, but this system of storing numbers can be extended indefinitely and therein lies its power. If we take two storage devices and consider them together, there are four possible conditions they can be in: 00, 01, 10, and 11. We can use these four pairs of symbols as numbers if we wish and we can enumerate four groups of objects with them from the least numberous, 00 (none), to the most numerous, 11 (three). Since this numbering system employs only two symbols, 0 and 1, it is called a *binary numbering system*. The numbering system we are most familiar with has ten symbols, 0 through 9, hence it is called a *decimal* numbering system. The binary system as we have developed it so far uses two bits of information stored on two separate devices, hence, it is called a *two-bit binary code*.

If we consider three storage devices together, we have a three-bit binary code with eight conditions, 000, 001, 010, 011, 100, 101, 110, and 111, corresponding to the decimal numbers 0 through 7. Extending the method to a four-bit binary code makes it possible to express the sixteen decimal numbers 0 through 15 as shown in Table 1-1.

Two or more bits of information constitute a *word*. Table 1-1 shows the sixteen possible 4-bit words representing, for the moment, the numbers 0–15. It should be evident that, with enough bits, we can represent any positive integer in binary code. Hence, any positive integer can be stored in computer memory, provided the memory has enough bistable devices. By allocating certain bits to locate the decimal point (strictly speaking, the

Table 1-1
Four-Bit Binary Numbers and Their Decimal Equivalents

| | | | | | | | |
|---|---|---|---|---|---|---|---|
| 0000 | 0 | 0100 | 4 | 1000 | 8 | 1100 | 12 |
| 0001 | 1 | 0101 | 5 | 1001 | 9 | 1101 | 13 |
| 0010 | 2 | 0110 | 6 | 1010 | 10 | 1110 | 14 |
| 0011 | 3 | 0111 | 7 | 1011 | 11 | 1111 | 15 |

radix point), and others to designate an exponent and a sign, decimal and exponential signed numbers can be stored in memory and our data storing capability becomes virtually limitless. Present-day computers commonly use 16 or 32 bit words (making possible $2^{16}$ or $2^{32}$ different combinations), but much larger words are possible. Contemporary microcomputers commonly use 8 bit words, although 16 bit microcomputers are commercially available. Because it is so common in microcomputer work, the 8 bit word is given a special name; it is called a *byte*.

The logical method used to generate Table 1-1 may be continued indefinitely, but evaluation of the decimal equivalent of binary numbers of five, six, or more bits can be tedious. Instead, we convert any binary number to its decimal equivalent by taking the following sum

$$y = \sum b\,(2)^{\mathbf{a}} \qquad (1\text{-}1)$$

where $b$ is either 0 or 1 and **a** is the *place number* in which the symbol is found, counting from right to left and starting with zero. Thus, $\mathbf{a} = 0$ for the rightmost place, 1 for the first place to the left of it, 2 for the place left of that and so on.

. . . 543210 Place numbers, **a**
. . . 101010 Binary number

The symbol $\Sigma$ indicates that the sum of all terms in Equation 1-1 is to be taken.

We can develop the first few entries in Table 1-1 from Equation 1-1 as follows:

$$\begin{aligned} y_0 &= 0 \cdot 2^3 + 0 \cdot 2^2 + 0 \cdot 2^1 + 0 \cdot 2^0 = 0 \\ y_1 &= 0 \cdot 2^3 + 0 \cdot 2^2 + 0 \cdot 2^1 + 1 \cdot 2^0 = 1 \\ y_2 &= 0 \cdot 2^3 + 0 \cdot 2^2 + 1 \cdot 2^1 + 0 \cdot 2^0 = 2 \\ y_3 &= 0 \cdot 2^3 + 0 \cdot 2^2 + 1 \cdot 2^1 + 1 \cdot 2^0 = 3 \end{aligned}$$

(Any number to the zero power is 1.)

The bit with the largest value of **a** (the leftmost bit) is called the *most significant bit* and the bit with $\mathbf{a} = 0$ is the *least significant bit*.

Computers can store not only numbers, but letters of the alphabet and symbols such as $ and ?, as the reader discovers each month when the bills come in. This is done purely by convention among computer designers. It is agreed that a certain binary word will be taken to represent a number in one context and the letter "A" in another. Another binary word may represent a number or "B" and so on. One such set of conventions is The American Standard Code for Information Interchange or ASCII convention. In which context a word of information is to be interpreted, *arithme-*

*tic* or *alphabetic*, is specified by the programmer in ways we shall encounter below.

### Exercise 1-1

Find the decimal equivalents of the following binary numbers.

(a) 1101
(b) 10101
(c) 110001
(d) 10010100

Solution 1-1. Remember that **a** increases according to the place number counted from right to left according to the series 0, 1, 2, 3, . . .

(a) $y_a = 1 \cdot 2^3 + 1 \cdot 2^2 + 0 \cdot 2^1 + 1 \cdot 2^0 = 8 + 4 + 0 + 1 = 13$
(b) $y_b = 1 \cdot 2^4 + 0 \cdot 2^3 + 1 \cdot 2^2 + 0 \cdot 2^1 + 1 \cdot 2^0 = 21$
(c) $y_c = 1 \cdot 2^5 + 1 \cdot 2^4 + 1 \cdot 2^0 = 49$
(d) $y_d = 1 \cdot 2^7 + 1 \cdot 2^4 + 1 \cdot 2^2 = 148$

## Data Processing

The ability to store and write out any amount of arithmetic and alphabetic data is not of much use unless we can perform arithmetic manipulations on such data. These manipulations are called *data processing*. Computers can add, subtract, multiply, divide, and perform logical operations.

Suppose we have stored a four and a two in computer memory and that the computer is wired so that if the detector senses a 1 in one of two memory locations, it sends an electrical pulse that changes the state of the corresponding bit of the other. If it senses a 0, it sends no pulse.

Checking with Table 1, we see that the leftmost number in Fig. 1-2 is a binary 2 and the middle one is a binary 4. If our detector–pulser scans the binary 2, it sends only one pulse, that corresponding to the second bit from the right. This changes the second bit from the right in the binary 4 from 0 to 1 resulting in the word 0110, a binary 6. Thus, we have performed the



Fig. 1-2. Adding by pulsing one location in computer memory according to the contents of another.

addition 2 + 4 = 6. With sufficient bits available, any two numbers can be added. In a similar way, the data processor is able to subtract, multiply, divide, and exponentiate. Subtraction is the opposite of addition. Multiplication is done by repeated addition, i.e., $A \cdot B$ is carried out by adding $A$ to itself $B$ times. Division is done by repeated subtraction and exponentiation is repeated multiplication, hence $X^3$ would be computed by taking $X \cdot X \cdot X$.

Logical operations are carried out in the processor by *logic gates*. Frequently, we wish to compare two numbers to find out whether they are the same. One way might be to use an AND gate and an OR gate.

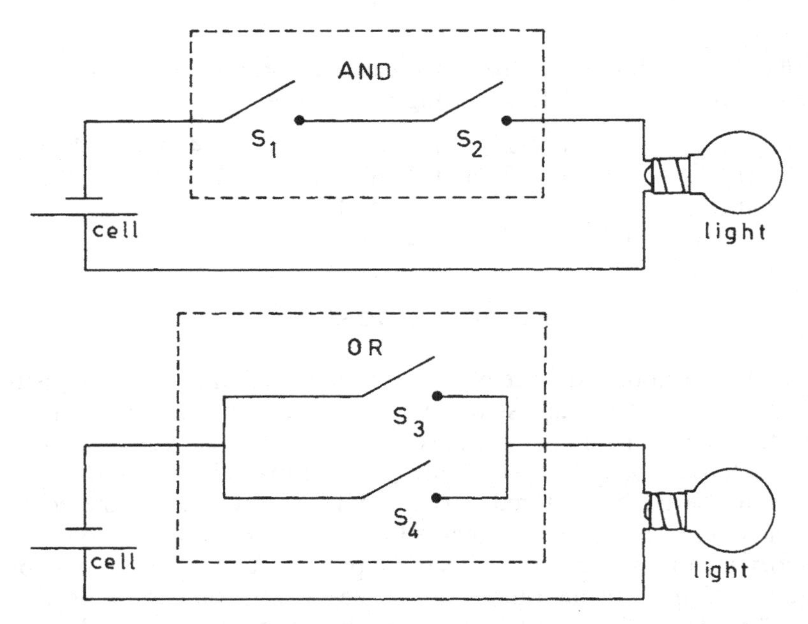


Fig. 1-3. The AND and the OR logic gates (real gates use transistors in place of electromechanical switches).

We can make statements about the conditions of switches $S_1$ and $S_2$ of an AND gate even though they may be obscured from our view and closed in a "black box" indicated by the dotted line. Since the battery, $S_1$, $S_2$, and a light bulb are connected in *series*, we know that if the light is on, $S_1$ *and* $S_2$ are closed. If the light is off, at least one switch must be open, possibly both. The OR gate is wired differently and we get different information from it. If we see that the light is on, we know that *either* $S_3$ *or* $S_4$ is closed, possibly both. If the light is off, both switches are open.

Let us compare the words 0100 and 0100 by means of logic gates to see whether they are the same. Let our detector scan both numbers bit by bit from left to right and transmit pulses to both the AND and the OR gates just described. If the detector finds a 1, it closes a switch, if it finds a zero, it leaves the switch alone. Beginning with all four switches open, the detector finds a 0 as the leftmost bit of the first number and transmits no pulse to $S_1$. It finds 0 as the leftmost bit in the second number and transmits no pulse to $S_2$. Both $S_1$ and $S_2$ stay open and the light is off (no output signal). Thus, we know that this first bit cannot be a 1 in the first number *and* 1 in the second. The same information transmitted to the OR gate leaves both switches, $S_3$ and $S_4$ open indicating that *neither* bit is 1, consequently, they are identical, both 0.

Moving one bit to the right, the detector senses 1 in both the first and second numbers, closes both switches of the AND gate and transmits a positive output signal (light), indicating that the bits are identical and both are 1. Moving one more bit to the right and reopening all closed switches, the results of the previous paragraph (0) are observed, as they are on the rightmost bit. Hence, the last two bits of each word are identical, and are zero. Since all bits in the two words are identical, the words are identical. Comparison is very important in computer decision making, which is at the heart of any program. The IF and FOR–NEXT statements, to be used presently, depend on decision making of this kind.

## Control

We have now explained, very briefly, ways of storing and manipulating numbers and letters, information input and output, and ways of making decisions. The final thing we need in a computer is a set of directions indicating the sequence in which these operations are to be carried out, i.e., a *program*. This also is done in binary code or some modification of it using only the symbols 0 and 1. By convention, certain binary words are taken to indicate instructions such as add, subtract, move to another location in memory, and so on. Instructions in this form are said to be in *machine language*, are stored in the memory of the computer, and are executed in the sequence in which they were entered unless directions to the contrary are written into the program. All large computers operate this way, hence they are called *stored program* computers.

Many people can and do communicate with computers directly in machine language. It is a very cumbersome language to use, however, and few nonspecialists are willing to put up with the trouble of writing any but the most simplistic programs in machine language. Moreover, the possibility of making a mistake while punching seemingly interminable strings of zeros and ones into a computer memory is very great. Only one symbol need be reversed or out of place, and, of course, the entire program refuses to run.

Consequently, it was thought that a new language could be made up consisting of groups of letters suggesting the operation to be carried out and that a special program be written to translate from this primitive alphanumeric language (language containing both letters and numbers) to even more primitive machine language. The new alphanumeric or mnemonic language was called *assembly language* and the program that converted programs witten in assembly to machine language was called the *assembler*.

Although the advent of assembly language was a big improvement, it did not resemble English much, and a program written in assembly language still looks an incomprehensible mess to the beginner. The next step, and the point at which it became practical for nonspecialists to begin to use computers with some degree of convenience, was the introduction of high-level languages. If machine translation of assembly language worked, then why not carry the process one step further and invent a language that would contain familiar words such as IF, GO TO, READ, etc., instead of the much less comprehensible assembly language? Each high-level statement may correspond to several assembly statements. Each assembly statement may contain more than one machine-language statement. Hence the high-level programmer calls down a whole throng of machine-language statements with a few English words. Of course, there has to be a program, more complicated than the assembler, to do all this. It is called the *compiler*. One might say that the compiler contains the assembler within it because it does everything the assembler did and more.

There are many high-level languages and many dialects. Probably the best known was designed for scientific problem solving, and, since it largely performed the function of scientific *for*mula *tran*slation, it was called FORTRAN. The language we shall use here is a very close relative of FORTRAN called BASIC. As its name implies, it is a little simpler and a little closer to the way that we formulate algebraic problems than FORTRAN. It suffers the disadvantage of being less flexible for very complicated and involved scientific problems. It has sufficient power to solve all of the problems we are interested in, however, and can be more easily learned and read. A more recent popular language is PASCAL. The reader should not fear having to begin all over if he or she wants to change languages. Unlike human languages, a second or third high-level computer language is very easy to learn.

## Kinds of Computers

Except for an occasional reference, this book will be concerned with general-purpose computers, which are frequently large and can be somewhat imposing machines. As the complexity of machines increases, "hands on operation" by students and even by faculty is becoming less common. This is unfortunate for it gives the computer an air of mystery that this book is attempting to dispel. Nevertheless, the trend is undeniable and will continue. Many readers may not even see their particular "Wizard of Oz," but only communicate with it by teletype.

The term general purpose computer invariably implies a stored program computer that accepts a program through an input device, compiles the program into machine language and, as directed by a *control command* (usually RUN), runs it. We have seen that binary words stored in computer memory can have many interpretations: program statements, memory locations, arithmetic and logical operations, alphabetic symbols and numbers—usually data. The context in which a binary word is taken is largely determined by the control commands and program statements. For example, the statement PRINT A, B, C causes the binary words stored in memory locations designated A, B, and C to be printed as decimal numbers, whereas PRINT 'A' would cause the letter A to be printed out.

Compilers have been expanded in another very convenient way. Certain numerical cross checks have been built into compilers along with alphabetic error messages (with which the beginning programmer rapidly becomes familiar). Hence, PRINT A. B, C would result in an error message because the comma between A and B has been mistakenly input as a period, resulting in an illegal address. These errors are called *bugs* and getting rid of them is called *debugging* the program.

Microcomputers are fascinating gadgets that will certainly come into much wider use in the near future. Their small size has made them particularly attractive for use as *dedicated* computers, that is, computers that have one specific job to do for one specific instrument and are often housed within it. Persons familiar with the computer output in modern clinical labs, or with computer output from gas-liquid chromatographs in the analytical laboratory will have seen two uses of dedicated computers. Because the programming of dedicated micros is frequently done by the manufacturer, they fall outside the scope of a book on computer programming for data processing.

Within the last few years several companies have developed general purpose microcomputers and have marketed them at prices so low that they will soon be as common as typewriters in educational and scientific institutions. Most microcomputers manufactured for the popular market feature BASIC in ROM which means that a separate program has been hard-wired into Read Only Memory that, unlike RAM, cannot be changed or erased by the programmer. (As the name implies, you cannot write information into a read only memory.) The BASIC in ROM chip interprets BASIC input instructions for the particular CPU or microprocessor chip used in that system and presents them to the CPU as they are being executed. This is in contrast to a compiler, which is a stored program that converts the entire input program into a machine language program called the *object program* stores it, then executes it all at once. A compiler is faster than an interpreter, but it also requires more memory.

Although there are no two systems that are completely language compatible, BASIC in ROM is essentially the same as the basic used in this book. Almost all of the programs you will find here have been run on microcomputers without significant change. One exception is the group of matrix statements in Chapter 11, which are not available in ROM packages at the time of this writing.

## Batch Processing and Time Sharing

In the past, it was usual to think of FORTRAN as a language that is batch processed and BASIC as a time-sharing language. Such restrictions are by no means valid because both languages can be processed both ways.

In batch processing, one punches the program into ordinary computer cards, punches a set of data cards that contain the numerical data to be worked up, and intersperses a number of control cards that indicate rudimentary control functions, such as which compiler is to be used, when to accept data, when to run the program, and when the job is done. The card decks are put into the computer and each deck is run sequentially. Results are normally printed out on an on-line printer.

One drawback of batch processing is that the lag between the time at which a program is given to the operator and the time when the output is returned. This lag is called the ''turnaround time'' and may be long for several reasons: a large number of programs with a higher priority than yours, a computer that is too slow for the jobs assigned to it, or indifference on the part of the operator.

When a time-sharing system is used with a sufficiently fast computer, this drawback should disappear. In time sharing, there is normally a number of computer terminals connected to the computer at one time. The computer receives input from each terminal in succession and replies to the input it receives. The computer spends only a fraction of a second with each terminal, but returns to it a few fractions of a second later to receive and act upon any new information that may be there. To the programmer, it seems as though his or hers is the only terminal on line and turn-around time approaches zero. The situation is like that in which a chess grandmaster, playing simultaneously with 15 opponents, circles the room making moves, and by the time you have thought of your reply to one move, has returned again with the next.

Computer time sharing is often accomplished through an *acoustical coupler* or *modem*, a device a little smaller than a shoe box that is fitted so that a telephone receiver can be placed firmly into it. One dials the number of the computer, which may be that of one's own computer or of one of the computer companies that rent out terminals on a pay-as-you-program basis. Upon establishing contact with the computer, a high-pitched tone is heard in the telephone receiver which, upon being placed into the acoustical coupler, establishes contact between the computer and the terminal. There is an appropriate log-in procedure designed to prevent use of the terminal by unauthorized people, after which the computer sends back some reply indicating that it is ready to receive your program. The details of log-in differ from one system to another, hence the instruction booklet or resident instructor should be consulted.

## When Not to Use the Computer

Although the computer does many things well, there are equally many things that it does poorly or not at all. It seems important to enumerate, at the outset, problems for which the computer is not well adapted and worth some time to dispel the layman's notion that "put it on the computer" is a scientific panacea in problem solving.

"One-shot" problems, that is, problems in which one computation is carried through to provide one answer, are usually not worth solving by computer. Thus, to take a trivial example, multiplying 639 by 42 could be solved by writing an appropriate program, but the time required would be much greater than the time necessary for a longhand computation.

However, if we had a thousand sequential numbers starting at $a_i$, and we wished to multiply each by the corresponding member of a thousand number sequential set starting at $b_i$, the use of a computer would be appropriate. The program to multiply $a_i \times b_i$ $(i = 1$ to $1000)$ would not take much longer to write than the program for $a \times b$ and, once written, would provide the thousand answers in a tenth of a second or less. Because *run time* is negligible in comparison to programming time, highly repetitive calculations are ideal computer applications. As we shall see, difficult one-shot problems can sometimes be broken down into a large number of easy but repetitive steps called iterations. In this situation, a computer tactic for solving the problem might be appropriate.

For reasons similar to those discussed above, computers are not well adapted to solving problems involving large amounts of input data that are different from one run to the next. In general, if the longhand or desk-top calculator solution to a problem takes an hour and inputing the data takes three-quarters of an hour, there is not much impulse to go to the computer lab, particularly if it is raining.

Never expect the computer to solve a problem that you cannot solve yourself. The computer is a machine and does only what you tell it to do. Hence, if you cannot direct yourself along the proper pathway to a solution of a problem, you cannot direct a machine along that path either.

There is an "in principle" implicit in the preceeding paragraph. If a programmer knows how to do all the steps in a calculation, it is certainly possible to do them twice, three times, or many times. If the solution to the problem requires that the same sequence of calculations be carried out many thousands or tens of thousands of times without error, the total computation time may be greater than a lifetime and a human problem solver is bound to make an error sooner or later anyway. We say that the problem can be solved *in principle* because one knows how to do it, but that it cannot be solved in practice. It is in such areas that the incredible speed and accuracy of the computer have vastly expanded the capabilities of scientists who once had to consign their *in principle* solutions to the wastebasket, correct though they may have been, but who can now solve them with an hour or two of time on a fast computer.

There is a middle ground between one-shot problems and problems possible to solve in principle, but impossible in practice. Most of the problems in this book lie in that middle ground, embracing problems that scientists can do but which we know they will not do because of the time expenditure involved. One example is curve-fitting by the least-squares procedure for nonlinear functions. There can be no doubt that this, or an equivalent, mathematical procedure is the correct way to draw curves best representing $x$–$y$ pairs of experimental points, and journal editors are beginning to demand it. Heretofore, the many hours or days of hand computation time demanded by even a moderately sized data set discouraged proper treatment of data and encouraged the guesswork method of drawing a line through points that "looks good". As we shall show in later chapters, computer routines can be set up to fit numerous different functions to indefinitely large data sets that are sufficiently general to be used over and over on different data sets resulting from entirely unrelated experimental investigations.

## Program 1-1

We would like to write a program in BASIC that will read data into computer memory, then locate, retrieve, and write it out by means of standard I/O peripherals. *The reader with access to a computer should by all means run each program as it is encountered in this book.*

```
10  REM PROGRAM TO READ AND WRITE DATA  VIA I/O PERIPHERALS
20  LET A=5
30  LET B=15
```

```
40  LET C=25
50  PRINT A,B,C
60  END

READY
runnh

 5              15              25


TIME:  0.17 SECS.

READY
```

Commentary on Program 1-1. In BASIC, computer statements are usually numbered in the order in which they are to be carried out. The seemingly obvious numbering scheme, 1, 2, 3, . . . is usually not used. Instead, the numbering shown, 10, 20, 30, . . . is used in order to permit insertion of statements between the statements already present in a program by selecting an intermediate number. For example, if we type a statement 25 anywhere in or after the program, it will be executed between statements 20 and 30.

The REM at the beginning of statement 10 indicates that it is a remark and is not acted upon by the computer. REM statements are used to give titles to the program or to designate and explain various modules within a long program. As such, they are not at all necessary to the program, but merely serve as conveniences for the (human) reader.

The LET statements, 20, 30, and 40 have a meaning that is evident from elementary algebra, but their function in BASIC is a little more subtle and should be borne in mind throughout this book. LET A = 5 takes the number 5, translates it into binary code and stores it in memory in a specific location labeled A (in binary, of course). Statements 30 and 40 perform similar operations on the numbers 15 and 25.

The statement PRINT A, B, C instructs the computer to search locations A, B, and C, determine what number is stored in each, translate it from binary to decimal and print out the result on an appropriate output device.

Since the computer has no judgment, it must be told that the program is at an END. After listing the program (command: LIS or LIST ) the computer replies READY if the programming is done on a terminal printer, as it was in Program 1-1. The RUN command tells the computer to run the program it has compiled and stored in memory. The command RUNNH tells the computer to run the program, but to leave no heading. Other control cards are necessary in batch processing. The assistants at your local computer facility should be consulted for details because they differ from one facility to the next. Do not be disappointed if your facility does not batch process BASIC; it is not universally used in batch mode.

After the run, the time used is printed out and the notification READY indicates that the computer is prepared to accept and act upon a new command.

## Program 1-2

```
10  REM PROGRAM TO CALCULATE THE SUM OF FIVE NUMBERS
20  READ A,B,C,D,E
30  DATA 25.1,19.3,24.4,21.5,18.8
40  LET S = A+B+C+D+E
50  PRINT S
60  END

READY
runnh

 109.1


TIME:  0.12 SECS.
```

## Modification 1-2A

```
10  REM PROGRAM TO CALCULATE THE SUM OF FIVE NUMBERS
20  READ A,B,C,D,E
30  DAtA 25.1,19.3,24.4,21.5,18.8
40  LET S = A+B+C+D+E
41  PRINT "THE SUM OF"
42  PRINT A,B,C,D,E,
43  PRINT "IS"
50  PRINT S
60  END

READY
runnh

THE SUM OF
 25.1           19.3           24.4           21.5           18.8
IS
 109.1


TIME:  0.19 SECS.
```

**Commentary on Program 1-2.** Program 1-2 shows a more convenient way to introduce data into a BASIC program in statements 20 and 30. Instead of using five LET statements, we have used two statements, READ and DATA. The input data are separated by commas. This is the first example of a principle to be developed throughout this book. The program is left open-ended and the reader should be able to see that it can be extended to sum any number of input data from two to 286. This upper limit is established by the usual rule that time-shared BASIC variables may only be a letter of the English language or the letter followed by a number. We shall look into ways of releasing this restriction below. Microcomputer BASIC is usually not as restricted as time-shared BASIC, but the rules vary from one system to another.

The LET S = . . . statement stores the appropriate sum in a location labeled S. The PRINT and END statements were encountered in the first program.

A very common method of programming is to write the bare essentials of a program and then add to it to obtain a more readable output. The modification of Program 1-2 shows a simple example. It is always a good idea to make at least a partial explanation of output data rather than to print out a number and leave the reader, who may not know BASIC, to figure out what it is. Statements 41 and 43 help to explain the modified output. These are also the first examples of alphanumeric output we have seen. Anything enclosed in quotes will be executed by the PRINT command exactly as it appears in the source program. Thus, PRINT 'XYZ 231 - ?A' produces the computer output XYZ 231 - ?A. The PRINT alphanumeric statement provides a convenient way to label tables and the like.

Statement 42 prints out the input data in just the way we saw in Program 1-1. It is always a good precaution to have the input data printed out, for it makes input errors easier to spot.

## Program 1-3

```
10  LET A=0.
20  FOR I=1 TO 10
30  LET A=A+1.
40  PRINT A
50  NEXT I
60  END

READY
RUNNH

 1
 2
 3
 4
 5
 6
 7
 8
 9
 10


TIME:   0.14 SECS.
```

**Commentary on Program 1-3.** Program 1-3 is designed to count from one to ten. It looks simpler than the previous two, but it is more subtle because it includes the first example of computer decision making and it makes use of a loop. Statements 20, 30, 40, and 50 constitute a for FOR-NEXT loop and it works like this.

After *initializing* the variable A at zero, we enter the loop *via* statement 20 which means "do everything in the loop for values of the running index I from 1 to 10." In other words, everything from statements 20 to 50 will be repeated ten times, or go through ten *iterations*.

Statement 30 *increments* the value of A in memory by adding 1 to it and storing the result, 0 + 1 = 1 in location A. The following statement prints A and the NEXT I statement answers the question "has this loop been executed ten times?" In other words, it compares the value I = 1 with 10, the required number of iterations, and finds that they are not the same (see the section on AND and OR gates.) This sends the program back to statement 20 with I now equal to 2 on the second iteration. A is incremented to 1 + 1 = 2 and printed, the NEXT I statement decides that the loop still has not been satisfied and the program goes into another iteration.

Finally, when I = 10, A is incremented to 9 + 1 = 10 and printed. The statement NEXT I determines that the loop has been satisfied and exits to the 60 END statement. A little thought should show that loops can be very powerful tools in more complicated programs.

## Glossary

*Acoustical Coupler.* Device permitting communication with a computer using an ordinary telephone line (sometimes called a *modem*).

*Alphanumeric Code.* Code containing numbers, letters, and symbols.

*Assembler.* Program used to translate a symbolic assembly language program into a machine language program prior to transmitting it to computer memory.

*Assembly Language.* Symbolic alphanumeric language for various simple computer operations.

*Batch Processing.* Sequential processing of programs, usually taken from cards.

*Binary Numbers.* Numbers using only two symbols, typically 0 and 1.

*Bistable Device.* Device that may exist in either one of only two states, e.g., a switch or a transistor.

*Bit.* A piece of information that can be represented in the binary number system as a 0 or a 1.

*Bug.* An error in a program.

*Compiler.* Program used to translate a high-level language program to a machine language program prior to transmitting it to computer memory.

*Computer Memory*. The part of a computer that stores binary coded instructions and data.

*Command*. Instruction to a computer, e.g., RUN, which is more general than a statement.

*Cores*. Ferromagnetic rings capable of storing one bit of information each, according to their direction of magnetization.

*Data Processing*. Arithmetic and other mathematical operations carried out on input data.

*Debug*. To find and correct all errors in a program.

*Decimal Numbers*. Numbers using ten symbols, typically the symbols 0 through 9.

*Dedicated Computer*. Computer built to perform only one or a limited number of functions; as contrasted to a general purpose computer.

*High-Level Language*. English word symbols calling up groups of assembly language statements for the purpose of executing mathematical operations or making logical decisions.

*Increment*. To increase a variable, usually by a set and constant amount.

*Initialize*. To set a variable at a specific initial value that may be changed as the program runs.

*Input Device*. Device permitting the user to transmit statements, commands, or data to a computer.

*Interpreter*. Program that converts high-level instructions into machine language as the program runs.

*Iteration*. One execution of a sequence of calculations that will be repeated many times during a program run; one execution of a loop.

*Least Significant Bit*. Rightmost bit in a simple binary word.

*Logic Gate*. Electronic element permitting current to flow only when a certain logical condition is satisfied.

*Loop*. Part of a computer program designed so that the machine goes back over it and repeats the calculations within the loop many, perhaps very many, times.

*Machine Language*. Information expressed entirely in terms of binary arithmetic.

*Modem*. Acoustical coupler.

*Most Significant Bit*. Leftmost bit in a simple binary word.

*Object Program*. Machine language program generated by the compiler.

*Output Device*. Device permitting the computer to transmit information, e.g., calculated results, to the user.

*Peripherals*. Parts of a computer, necessary to its useful functioning but not part of the central processing unit.

*RAM*. Random Access Memory.

*Run Time*. Amount of time the computer takes to execute a program exclusive of compilation time.

*Source Program*. Program requiring compilation or interpretation that gives the series of algebraic and logical steps a computer must go through to solve a problem.

*Stored Program.* Sequence of directions stored in computer memory and capable of being carried out at some later time on input data.

*Statement.* Instruction to a computer specific to one program, e.g., PRINT X.

*Subroutine.* Series of program steps called up by one or a limited number of statements smaller in number than itself.

*Time Sharing.* System by which several users may use a computer (essentially) simultaneously, usually using teletype terminals.

*Turnaround Time.* Time interval between submission of a program and receipt of computed results.

*Word.* A collection of bits.

## Problems

*1.* Find the decimal equivalents of the following binary numbers:

A. 1010
B. 01101
C. 101011
D. 1110111

*2.* Add 3 and 6 in four-bit binary arithmetic.

*3.* Write a paragraph describing what happens when the binary numbers 0100 and 0110 are compared using an AND and an OR gate.

*4.* If the memory locations in a microcomputer are specified by a byte (8 bit word), how many memory locations can be addressed? If memory locations or addresses are specified by two bytes, often called a HI byte and a LO byte, how many locations can be addressed, i.e., how many distinct memory addresses are possible?

*5.* Write a program to multiply five numbers. (The symbol for multiplication is *.)

*6.* Write a program to add four numbers and divide the sum by 2. (The symbol for division is /.) Check your result with pen and paper. Did you run into any trouble? Can you locate its source and remedy it?

*7.* Write a program to count backwards from 10 to 1.

## Bibliography

R. L. Albrecht, L. P. Finkel, and J. R. Brown, *Basic,* Wiley, New York, 1973.

T. H. Crowley, *Understanding Computers,* McGraw-Hill, New York, 1967.

C. B. Dawson and T. C. Wool, *From Bits to Ifs,* Harper and Row, New York, 1968.

C. K. Mann, T. J. Vickers, and W. M. Gulick, *Basic Concepts in Electronic Instrumentation,* Harper and Row, New York, 1974.

S. P. Perone and D. O. Jones, *Digital Computers in Scientific Instrumentation: Applications to Chemistry,* McGraw-Hill, New York, 1973.

# Chapter 2

# Understanding Experimental Error and Averages

The numbers we use to represent experimental data express more than just the magnitude of a set of measurements. They also express the experimenter's best estimate of the accuracy of the measurements. Thus, if we say a steel rod is 27 cm long, we are not committing ourselves to the degree of accuracy that we are if we say that it is 27.032 cm long. By one convention, the last digit is taken to be an uncertain digit; hence, 27 cm implies that the measurement is taken to be about 7/10 of the way between 20 and 30 cm, while the datum 27.032 implies that it is about 2/10 of the way between 27.03 and 27.04 cm.

Another way of expressing this measurement might be to say that it is 27 ± 1 cm long (27 plus or minus 1 cm) indicating that the error is not thought to be more than 1 cm either way. This can only be a guess, for if we had any way of knowing exactly what the error is, we could correct for it arithmetically and write down an error-free measurement. Hence, in the use of significant figures or in writing down an estimated error interval, there is always a probability that we are wrong and that the actual error falls outside the specified range. The method by which we arrive at our estimated error range determines how large this probability is, but for a measurement on a continuous variable, e.g., length, it is never zero.

The impossibility of an error-free measurement of a continuous variable is illustrated by taking our measurement of 27.032 cm and improving our techniques such that we obtain the value 27.0324. Better, but still the measurement tells us only that the length of the steel rod is about 4/10 of the way between 27.032 and 27.033 cm. Improving our measuring techniques again to obtain 27.03241 leads to the same dilemma, and so on *ad infinitum*. Hence, for this kind of measurement, there is no correct answer, but (classically) we can reduce our error as much as time, patience, and apparatus permit.

By contrast to continuous variable measurements, there are also integer variable measurements that can be error-free. Clearly, it is possible to say unequivocally that there are three people in a room.

## Significant Figures

No amount of computation can ever make data derived from a single experimental measurement more accurate than the experimental measurement was to begin with. Moreover, when a result is calculated from a number of experimental measurements, the result can never be more accurate than the least accurate of the measurements. To avoid implying a greater or smaller degree of accuracy in a computed result than the data warrant, a few rules of arithmetic manipulation are used.

*1. When two numbers are added, the smallest number of digits beyond the decimal point is retained in the sum.* Thus,

$$37.57 + 3.932 = 41.50$$

not 41.502. Confusion arises when one wishes to add numbers like 72,000 + 6,590. Are the zeros to the right of these numbers significant or do they merely serve to locate the decimal point? The ambiguity is removed by expressing all numbers, in accepted scientific style, as exponential numbers. Thus, $7.2 \times 10^4$ has two significant figures and $7.200 \times 10^4$ has four. Now there is no ambiguity in writing

$$7.2 \times 10^4 + 6.59 \times 10^3 = 7.9 \times 10^4$$

and

$$7.200 \times 10^4 + 6.59 \times 10^3 = 7.859 \times 10^4$$

*2. When multiplication is carried out, retain the same number of significant figures as the least accurate measurement.* Thus,

$$32.983 \times 4.2 \times 0.99982 = 1.4 \times 10^2$$

after rounding back to the proper number of significant digits. This rather disheartening result suggests that we had better find a more accurate way of making our middle measurement, for we are throwing away most of the accuracy inherent in the other two. The exception occurs when one or more values input to a computation is an integer, whereupon it may be assumed to have as many significant zeros beyond the decimal point as we choose. Thus,

$$39.983 \times 4 = 159.93$$

provided that 4 is an integer and not a measured continuous variable.

3. *In rounding, if the first digit beyond the last significant digit is less than 5, drop all insignificant digits; if it is more than 5, increase the last significant digit by one and drop the rest*. Thus, rounding to three significant digits, 12.34 becomes 12.3 and 12.36 becomes 12.4. The only ambiguous case arises when the digit to be dropped is exactly 5. Some scheme of alternation or splitting the difference is usually used. One way is to raise when the digit preceding the five is not even and not to raise it when it is odd. Since numerous rules possible here are equally arbitrary, suit yourself as to choice, or make up your own, but be consistent about the use of the rule you choose.

Ambiguities do arise when using the significant figure conventions, particularly in multiplying data having different numbers of significant figures and in taking their logarithms and antilogarithms (see Skoog and West, 1980). There are unambiguous ways of expressing experimental uncertainty to be discussed in the next chapter and developed throughout this book.

## Sources of Errors

The three principal kinds of errors in experimental measurements are *systematic errors, random errors,* and *blunders*. The first two kinds are present in all measurements of a continuous variable, and the last should be very rare.

Systematic errors usually owe to faulty instrument calibration, a simple example being a rifle that has not been properly "sighted-in." No matter how good a marksman may be, several rounds are unlikely to be grouped in the bulls-eye without sighting-in the gun first or making some other compensatory adjustment, such as shooting high, to the left, and so on. Scientific instruments must usually be calibrated against some known standard to avoid systematic errors. A common example is calibration of the wavelength output of a spectrophotometer using a sample with a sharp absorption peak at known wavelength. With proper calibration methods, systematic errors can usually be brought down to a negligible part of the intrinsic resolution of the measuring instrument.

There are, however, always indeterminate errors associated with experimental measurements. They may be reduced by refining experimental technique, but they may never, because of the nature of continuous variables, be made zero. These errors are usually assumed to be random in nature and if, for example, we are measuring the length of a steel rod, we assume that our measurement is just as likely to be too short as it is to be too long.

Completely random errors often follow a *normal* distribution and will be the subject of considerable discussion in later chapters. Errors may be normally distributed in one dimension, as in the measurement of a steel rod, in two dimensions, as in firing at a target, or in many dimensions.

Blunders should be extremely rare when trained researchers work in a properly-organized laboratory, but they do occur. Gross misreading of an instrument, mislabeling of a reagent, and so on constitute blunders. Even though undetected at the time, they can frequently be detected later by statistical treatment of the data, and rules will be discussed below for throwing out experimental data on statistical grounds. The best procedure to follow when a blunder is suspected is to repeat the entire experiment, but that is, of course, not always possible.

*Accuracy* is the degree of agreement of a series of experimental measurements with the ''true'' value of the quantity measured. In some areas, such as calibration of an instrument against a known standard, we have a value which we can regard as true—one that has been measured many times using a procedure more reliable, or an instrument inherently superior to the one we are using, or both. In the general experimental situation, however, we do not have a known value, for if we had it we would not be making the measurement in the first place. In such situations, we may guess the accuracy from measurements made on a known system that is as nearly analogous to our system as possible. Alternatively, we may guess the accuracy of a set of data from its precision.

*Precision* is the amount of ''scatter'' or deviation of a set of experimental data away from some measure of central tendency such as the mean or average discussed below. The concept of precision does not require some hypothetical known or true value for comparison. Thus in a real experimental situation, the precision can always be measured, while the accuracy usually can not. If the experimental apparatus and procedure have been checked with sufficient thoroughness, it is common to take the precision as an estimate of the accuracy. This amounts to assuming that the systematic error is zero or negligible with respect to a normally distributed random error.

## Truncation Errors

No matter how large the word length stored in computer memory, some bits are inevitably lost. Suppose we are considering a 16 bit word length. When a datum is input (generally in decimal form) it is translated to binary arithmetic and stored in memory before being processed in the CPU, which also handles a specific number of bits. In some cases, all the binary bits will fit into the 16 bit memory location, but usually there are too many. The 16 most significant bits are stored and the less significant bits (those to the right of the stored sixteen bits) are lost. After processing, the result is stored in a 16 bit register and ultimately translated to decimal arithmetic and printed out.

Usually sixteen bits are more than enough to accommodate the accuracy we need for simple data processing; hence, the error that results from cutting off or *truncating* a stored number is negligible. There are, however, situations in which it is not. Small machines such as calculators and microcomputers usually give slightly inaccurate answers when very large or very small exponential numbers are computed.

There are also situations in which truncation errors are serious. Consider the attempt to determine the value of $e^{-x}$ by computing the sum $1 - x/1! + x^2/2! - x^3/3! + x^4/4! - \ldots$ to which it is equal. We know that $e^{-x}$ where $x = 10$, for example, is a very small number, but evaluation of any of the higher terms in the sum, e.g., $x^4/4!$ is a large number. Evidently $e^{-x}$ is the sum of large terms of alternating signs that almost, but not quite, cancel each other. A tiny truncation error in any of the terms of the sum is carried over into a serious error in $e^{-x}$. It is as if we were weighing an elephant with a flea on its back and the same elephant without the flea in an attempt to determine the weight of the flea by difference. A minute error in either weighing gives an absurd result for the weight of the flea.

Because of this, small discrepancies between results in this book and those calculated at other computing centers using different computer hardware can be expected, particularly in Chapter 11. But there should be no significant differences.

## Propagation of Errors

Since experimental data are processed in order to obtain a calculated result, errors made while taking the measurements are also processed and lead to an error in the result. We would like to get some general idea of the way errors are propagated, and of the nature of the computational factors that determine their ultimate size.

Consider measurement of the volume, $V$, of a cylinder

$$V = \pi r^2 h \tag{2-1}$$

where $r$ is its radius and $h$ is its height, both of which are input data for the computation. Measurements of both $r$ and $h$ suffer errors, $\Delta r$ and $\Delta h$, respectively, which may be positive or negative. Substituting the true values of $r$ and $h$ plus the measurement errors for each into Eq. (2-1) leads to

$$V + \Delta V = \pi(r + \Delta r)^2 (h + \Delta h) \tag{2-2}$$

where $\Delta V$ is the error in the computed result arising from $\Delta r$ and $\Delta h$. Expanding Eq. (2-2) algebraically,

$$V + \Delta V = \pi(r^2 + 2r\Delta r + \Delta r^2)(h + \Delta h)$$
$$= \pi(r^2h + 2hr\Delta r + h\Delta r^2 + r^2\Delta h + 2r\Delta r\Delta h + \Delta h\Delta r^2)$$

The first term in this sum leads to $\pi r^2h$, which is equal to $V$ by Eq. (2-1); hence, we can subtract $V$ from both sides with minor rearrangement of the terms to obtain the error,

$$\Delta V = \pi(r^2\Delta h + 2rh\Delta r + \Delta r^2h + 2r\Delta r\Delta h + \Delta r^2\Delta h) \tag{2-3}$$

In the normal laboratory situation of this kind, the error is only a few tenths of a percent, or, at most, a few percent of the quantity being measured. Thus, terms in Eq. (2-3) containing the square of an error or the product of two errors are very much smaller than terms containing only one or the other error (see problems). The value of this sum is not significantly affected by the presence of the last three terms, hence they can be dropped, which leads to

$$\Delta V \cong \pi(r^2\Delta h + 2rh\Delta r) \tag{2-4}$$

where the sign $\cong$ means "is approximately equal to."

Figure 2-1 shows the contribution of the two terms comprising $\Delta V$. The $\pi r^2\Delta h$ term contributes an error to the final measurement that would physically resemble a wafer (strictly speaking, a cylinder of small height $\Delta h$) that is added to the true volume and contributes to $\Delta V$. The $2\pi rh\Delta r$ term contributes a cylindrical shell to the error, which is of thickness $\Delta r$.

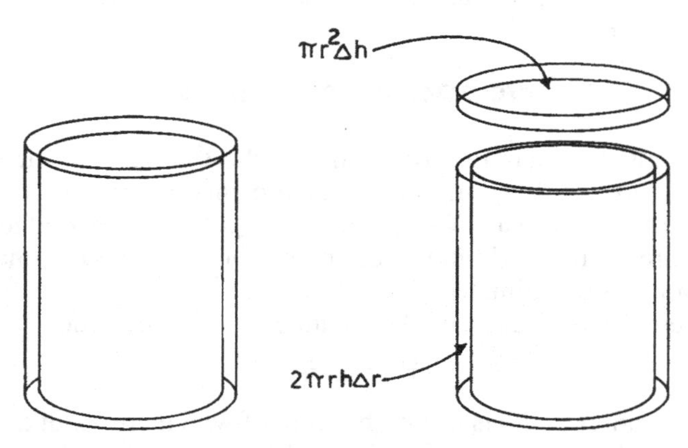


Fig. 2.1. Effect of errors $\Delta r$ and $\Delta h$ on the error $\Delta V$ of the computed volume of a cylinder.

As illustrated, both terms are positive, making the computed volume too large by a positive error, $\Delta V$. Both could be negative, making the computed value of the volume too small, or one could be positive and one negative, tending to cancel each other. The computed error could tend toward zero in this circumstance.

The fractional error $\Delta V/V$ is often a useful value to have. In the present case, it is

$$\begin{aligned} \Delta V/V &= \pi(r^2\Delta h + 2rh\Delta r)/\pi(r^2 h) \\ &= \Delta h/h + 2\Delta r/r \end{aligned} \tag{2-5}$$

## Exercise 2-1

Error analysis of the kind just discussed is frequently useful in experimental design. Suppose it is necessary to measure the volume of each of a set of cylindrical reaction chambers having a height of approximately 12 cm and a radius of approximately 3 cm. The nature of the experiment makes it necessary to know the volume of each reaction chamber with an error of 1.0% or less. Suppose further that we have determined our measurements of $h$, the easier of the two necessary measurements to make, to have an error of 0.010 cm or less. As a part of our experimental design, we need to know how accurate our measuring technique for $r$ must be, i.e., we want to know the upper limit on $\Delta r$.

Solution 2-1. The approximate volume of each cylinder is

$$V = \pi r^2 h \cong 340 \text{ cm}^3$$

One percent of this volume is 3.4 cm$^3$ and $\Delta V/V = 0.010$. Substituting other pertinent measurements, and the upper limit on the uncertainty in $h$, into Eq. (2-5),

$$\begin{aligned} \Delta V/V &= 0.010 = 0.010/12 + 2\Delta r/3 \\ 2\Delta r/3 &= 0.010 - 0.000833 = 0.00916 \\ \Delta r &= 3(0.00916)/2 \cong 0.0137 = 0.014 \end{aligned}$$

It is good practice to carry more than the appropriate number of significant figures through a calculation and round the answer to the proper number of digits at the end. The number of significant digits is slightly ambiguous in this case because the items 3 cm and 12 cm (above) are approximations.

This analysis tells us that, in order to stay within the required 1.0% error limit, the measuring technique for $r$ must be almost as good as that for $h$. To check our results, we substitute the upper limits on $\Delta h$ and $\Delta r$ into

Eq. (2-4) and obtain

$$\Delta V \cong \pi\ [3^2(0.010) + 2\ (3)\ 12\ (0.014)]$$
$$= \pi\ (0.090 + 1.008) = 3.4\ \text{cm}^3$$

which is the absolute value of the upper limit on the error in the volume as stated in the previous paragraph. It is interesting to note that the error in $r$ contributes about ten times as much to $\Delta v$ as the error in $h$. Hence, an improvement in the $r$ measurement reduces $\Delta V$ much more than an equivalent improvement in measuring $H$. This is a result that might not have been evident from a cursory examination of the demands made by the experimental design.

## Program 2-1

We wish to write a program to investigate both positive and negative errors in measurement of $r$ and $h$ of a cylinder. Taking the cylinder just described, with true values $r = 3.00$ cm and $h = 12.00$ cm, calculate the values of $V$ for $r = 2.50$ to $3.50$ cm in intervals of 0.10 cm holding $h$ constant at 12.00.

```
10 LET H=12.0
20 FOR R=2.5 TO 3.5 STEP 0.1
30 LET V=3.14159*R*R*H
40 PRINT V
50 NEXT R
60 END

READY
RUNNH

 235.619
 254.846
 274.826
 295.561
 317.049
 339.292
 362.288
 386.039
 410.543
 435.801
 461.814


TIME:   0.24 SECS.

READY
```

**Commentary on Program 2-1.** The only new instruction in this program is a way to control a FOR-NEXT loop using a variable rather than a running index as was done in Program 1-3. The loop starts with $R = 2.5$, statement 30 is a BASIC translation of the equation $V = \pi r^2 h$, and the remaining statements have already been discussed in other programs. When NEXT $R$ is encountered, control goes back to statement 20 and $R$ is incremented by the STEP size indicated, in this case, 0.1. If no step size is indicated, it is automatically taken as 1.0, as it was in Program 1-3. The output of Program 2-1 constitutes the eleven values of $V$ calculated for $r = 2.5$, 2.6, . . ., 3.5.

### Exercise 2-2

Change statement 10 in Program 2-1 to LET $H = 11.5$ and run it. Change 10 to LET $H = 12.5$ and run it again. Plot the results on the same graph paper along with the results you have already obtained at $H = 12.0$. Describe the three curves you have drawn.

Solution 2-2. If the curves are drawn with sufficient care, they will be seen to be three gently curving, equidistant parabolic arcs. In fact, they are contours of a very narrow slice of the parabolic sheet depicted in Fig. 2-2.

## Euler's Theorem and Variations

We shall use calculus to show how our method of splitting up the error or *variation* in $V$ resulting from simultaneous errors or variations in $r$ and $h$ leads to a general and powerful theorem of variations known as Euler's theorem. Taking the derivative of Eq. (2-1) with respect to $h$ at constant $r$ leads to

$$dV/dh = \pi r^2 \tag{2-6}$$

and the corresponding derivative with respect to $r$ at constant $h$ is

$$dV/dr = 2\pi rh \tag{2-7}$$

The reader should recall that quantities such as $dr$ are called infinitesimals, and that as a variation, $\Delta r$, becomes smaller and smaller, it approaches $dr$ in the limit.

If we multiply both sides of Eqs. (2-6) and (2-7) by $dh$ and $dr$, respectively, and add the resulting products, we get

$$(dV/dh)dh + (dV/dr)dr = \pi r^2 dh + 2\pi rhdr \tag{2-8}$$

in which the right side is identical to the right side of Eq. (2-4), except that it is written in terms of infinitesimals $dh$ and $dr$ instead of finite variations,

$\Delta h$ and $\Delta r$. Having arrived at the same result, we see that our method of taking derivatives is equivalent to the error analysis that preceeds it.

By analogy to Eq. (2-4), we may write the left hand side of Eq. (2-8) as an infinitesimal *total* variation, $dV$

$$dV = \left(\frac{\partial V}{\partial h}\right)_r dh = \left(\frac{\partial V}{\partial r}\right)_h dr = \pi r^2 dh + 2\pi rhdr \qquad (2\text{-}9)$$

which is a special case of Euler's theorem. We have used some changes in notation to indicate just what we did mathematically when we evaluated the infinitesimal variation in $V$ for simultaneous variations in $h$ and $r$. Although we knew that the volume depends upon two dimensions, $h$ and $r$, we took the derivatives independently, regarding $r$ as constant when we took the derivative with respect to $h$ and $h$ as constant when we took the derivative with respect to $r$. This is called a *partial differentiation* and is denoted, e.g., $\partial V/\partial h$ (read "the partial of $V$ with respect to $h$"). The notation $(\partial V/\partial h)_r$ is often used to emphasize that $r$ is held constant when the partial of $V$ is taken with respect to $h$. A similar notation is used for the partial of $V$ with respect to $r$. The general case of Euler's theorem says that if any variable, $Y$, is a function of any number, $n$, of independent variables, $x_1, x_2, \ldots x_n$,

$$Y = f(x_1, x_2, \ldots x_n)$$

the total variation of $Y$ for simultaneous variations in all the $x$ values is given by a sum analogous to Eq. (2-9),

$$dY = \sum_{i=1}^{i=n} (\partial V/\partial x_i)_{x_j} dx_i \qquad (2\text{-}10)$$

where $\Sigma$ denotes a summation, $i$ is the *running index* of $x$, $i = 1, 2, \ldots n$, and $x_j$ is any $x$ that is not $x_i$ (see below for more on summations).

## Geometrical Interpretation of Euler's Theorem

Retuning to our specific example of the volume of a cylinder, which is a function of its height and radius,

$$V = f(h,r)$$

we can get a better feeling for the significance of each term in Euler's theorem

$$dV = (\partial V/\partial h)_r\, dh + (\partial V/\partial r)_h\, dr \qquad (2\text{-}11)$$

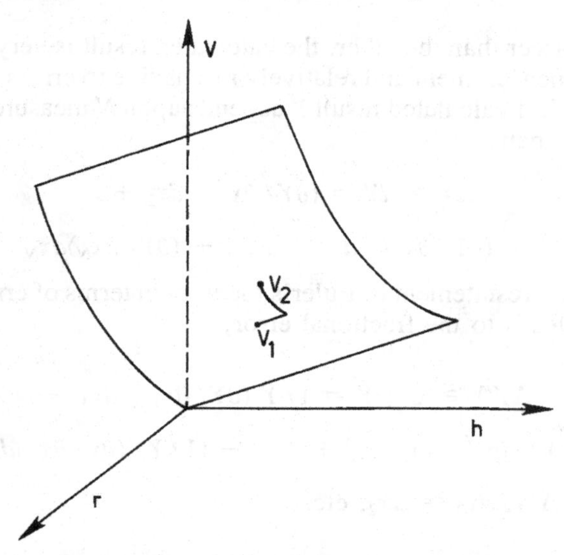


Fig. 2.2. Two dimensional surface representing $V = f(h,r)$. The surface is a parabolic sheet coming out of the page and tilted upward along the $h$ axis.

by imagining a curved surface in three-dimensional space, one horizontal dimension representing $h$, one horizontal dimension representing $r$, and the vertical dimension representing $V$.

From Eq. (2-1) we know that each and every pair of values of $h$ and $r$ produces a value of $V$, hence, $V = f(h,r)$ is represented by a surface. From any point, $V_1$, we may move to point $V_2$, infinitesimally distant, on the surface by an infinite number of different paths, but $dV$, the change in $V$, will be the same no matter which path we choose. This is equivalent to the truism that, no matter what route we choose from the bottom to the top of a mountain, our change in altitude is always the same provided that we start and finish at the same points.

Viewed geometrically, Euler's theorem notes that, of the infinite number of paths available to us, making part of the traverse at constant $r$, the first term in Eq. (2-11), and the second part of the traverse at constant $h$, the second term in Eq. (2-11), is the path easiest to analyze mathematically. Moreover, since all paths are equivalent, the result calculated from Euler's theorem applies to *any variation in* h *and* r, simultaneous or sequential.

The partial differential terms are a measure of the sensitivity of the variation of $V$ to small variations in $h$ and $r$ and are not, in general, the same. These sensitivity coefficients may be positive or negative. When one is much larger than the other, the calculated result is very sensitive to errors in one measurement and relatively insensitive to errors in the other.

In general, if calculated result $Y$ depends upon $N$ measured variables, $x_1, x_2, \ldots x_N$, then

$$\Delta Y \cong dY = (\partial Y/\partial x_1)_{x_j} dx_1 + (\partial Y/\partial x_2)_{x_j} dx_2 + \ldots + (\partial Y/\partial x_N) dx_N \quad (2\text{-}12)$$

This is merely a restatement of Euler's theorem in terms of errors or variations, which leads to the fractional error,

$$\Delta Y/Y \cong dY/Y = 1/Y (\partial Y/\partial x_1)_{x_j} dx_1 + (1/Y)(\partial y/\partial x_2)_{x_2} dx_2 + \ldots + (1/Y)(\partial y/\partial x_N) dx_N \quad (2\text{-}13)$$

where $dx_1 \cong \Delta x_1$, $dx_2 \cong \Delta x_2$, etc.

## Measures of Central Tendency

If we allow a marble to roll down a chute onto a flat table, then measure the distance it rolls, chance will play an important part in the results. Mechanistically, we might postulate a theoretically perfect marble, rolling down a similarly perfect chute and always winding up in the same spot. Our real observations will no doubt be quite different. Minor variations in starting position of the marble and minor irregularities in the chute or table will cause some variation in distance from one roll to another.

If we perform the marble-rolling experiment several times, there will be an interval that includes all measurements, say 90 to 100 cm. This interval is called the *range*, and may be made smaller by increasing the controls on the experimental conditions, but can never be made equal to zero. The size of range is determined by subtracting the smallest measurement from the largest, 100 − 90 = 10 cm in the example above. As we perform many more repetitions of the marble-rolling experiment, we notice that the distances rolled tend to cluster about some point in the range. Intuitively, we expect an "average roll" to be more representative of the group than any randomly selected individual roll.

The term "average" has been much abused in popular usage, and we shall generally avoid it in favor of more rigorously defined terms, such as the arithmetic mean, the quadratic mean, and so on. The arithmetic mean is the measure of central tendency that is usually meant by the term average, and it is by far the most important measure of central tendency used in scientific work. For the time being, we shall assume that we are working with *ungrouped data,* that is, data that appear in random order, not arranged into any particular set of qualitatively or quantitatively different groups. "Raw" data, just as it is collected in the laboratory or in the field, is usually ungrouped.

The arithmetic mean, $\bar{x}$, of a set of $N$ measurements is given by

$$\bar{x} = \sum_{i=1}^{i=N} x_i \Big/ N \tag{2-14}$$

where Greek sigma, $\Sigma$, indicates summation and the limits of the summation are given by the notation above and below the summation sign. Thus, $\sum_{i=1}^{i=N}$ indicates that all $N$ members of the set of values $x_i$ should be included in the sum

$$\sum_{i=1}^{i=N} x_i = x_1 + x_2 + \ldots + x_N$$

It is usually fairly obvious what values of $i$ are to be considered, as it is in the case of arithmetic mean, where $i$ can hardly do other than run from 1 to $N$. When the behavior of the running index is evident from context, its limits are not explicitly stated. Thus the arithmetic mean is usually written simply

$$\bar{x} = \sum x_i \Big/ \mathrm{N} \tag{2-15}$$

The *weighted arithmetic mean* arises when there is reason to suppose that some measurements in a set are more reliable than others. To illustrate, suppose we have two estimates, given by two different witnesses, of the age of a man seen running from the scene of a crime. The first witness says the man was 30 years old and the second says he was 40. The arithmetic mean estimate is (30 + 40)/2 = 35 yr. Subsequent interviews with the two witnesses show that the first witness tends to score poorly when asked to estimate the age of individuals in photographs, while the second tends to score consistently well. We might wish to reflect this performance in estimating the true age of the suspect but we do not wish to disregard the evidence of the first witness entirely. Hence we might arbitrarily give the evidence from the second witness twice the weight of that from the first and estimate the suspect's age as the weighted arithmetic mean

$$[30 + 2(40)] \Big/ 3 = 37 \text{ yr}$$

when rounded to two significant figures.

Notice that the sum of the weighting factors appears in the denominator. This procedure amounts to counting the evidence of the first witness, counting that of the second witness, and then counting that of the second witness again, as if there had been three witnesses, two of whom agreed. The denominator must be the number of effective witnesses counted, namely three. In the general case of many experimental measurements with many (perhaps equally many) dissimilar weighting factors

$$\bar{x} = W_1x_1 + W_2x_2 + \ . \ . \ . + W_Nx_N) \Big/ (W_1 + W_2 + \ . \ . \ . + W_N)$$

$$\bar{x} = \sum W_ix_i \Big/ \sum W_i \qquad (2\text{-}16)$$

The range serves as a rough measure of central tendency for, if we know that a set of data falls in a range from 100 to 150, we may suppose that the data tend to cluster somewhere in between. This presupposes a fairly large data set. The small, three-number data set, 100, 125, and 150, has a range of 100–150, but shows no tendency to cluster. In ensuing discussions, we shall always assume an indefinitely large data set unless otherwise stipulated.

The range is not a particularly good measure of central tendency because it is strongly influenced by one or two measurements that may differ greatly from the rest. It is not difficult to imagine a data set which clusters about 110, but has a range of 100–150. Such a data set would not be well represented by its range.

If the data are arranged in a regularly increasing order from lowest to highest (or in a regularly decreasing order from highest to lowest), the arrangement is called an *array*. The *median* is the point in an array that divides the data into two equal groups. In a data set containing an odd number of points, such as 1, 2, and 3, the median is a member of the set. In this case, the median of the set is 2. The median of an even numbered data set may not be a member of the set, e.g., the median of 1, 2, 3, and 4 is 2.5. Although the illustrations above used integer variables, the median of nonintegral variables is computed in the same way. From this point on we shall use the term "real variable" to indicate a value drawn from the real number set. Real or nonintegral data are usually, but not always, the results of measurements of a continuous variable.

The *mode* is the member of a data set that appears most frequently. In a small data set of integral variables there may not be a mode. The integers one through ten do not have a mode; each integer is represented only once.

A problem arises in calculating the mode of a real variable. Strictly speaking, no two measurements of a real variable are the same. If carried out to a sufficient degree of accuracy, two adjacent measurements will always be found to differ in the last significant digit. We therefore select an *interval* or subdivision of the range covered by the real variable. For the purposes of selecting the mode (and other purposes), all real variables falling within any interval are lumped together. Hence the measurements

8.92, 8.87, and 8.95 all fall within the interval 8.90–9.00, and if no other interval contains three or more numbers of the set, its midpoint, 8.95, is the mode of the set. When we come to predict future data distributions from past observations, we shall predict that, all other things being equal, the mode of past observations is the *most probable* of future events.

The *quadratic mean* or *root mean square* is widely used in certain physical applications.

$$\bar{x}_{rms} = \sqrt{\left(\sum x_i^{\,2} / N\right)} \qquad i = 1,\ldots,N \tag{2-17}$$

It is important in the theory of molecular speeds and, as we shall see in the next chapter, the quadratic mean of the deviations of a data set from some measure of central tendency is an important statistical quantity called the standard deviation. Note that the mean of the squares of a number set is not the square of the mean,

$$\overline{x^2} \neq (\bar{x})^2$$

## Exercise 2-3

Show that the mean of the squares of 2, 3, and 4 is not the square of their mean.

Solution 2-3. The mean of the squares of 2, 3, and 4 is

$$(4 + 9 + 16)/3 = 9.666.\ldots$$

The square of the mean is

$$[(2 + 3 + 4)/3]^2 = 9.000.\ldots$$

## Program 2-2

Program 2-2 is based on Example 2-3. We wish to write a program that will read in five values of $x$ and compute both the sum of the squares and the square of the sum of all five using a logical structure that is similar to that of Program 1-2. The logical structure of programs is independent of the language in which they are written. For the purpose of these commentaries, we shall introduce three terms descriptive of programs: the *strategy* of a program, which is the overall concept by which a given problem or problem type is attacked, the *tactics* of moving and manipulating blocks of data within a program in such a way as to contribute to the strategic solution, and the step-by-step *details* of writing the actual program. By and large, the first two are independent of the programming language, although the third is highly dependent upon language.

```
10  READ X1,X2,X3,X4,X5
20  DATA 23.5,41.8,12.3,42.1,63.55
30  LET S1=X1**2+X2**2+X3**2+X4**2+X5**2
40  LET S2=(X1+X2+X3+X4+X5)**2
50  PRINT X1,X2,X3,X4,X5
60  PRINT S1,S2
70  END

READY
RUNNH


 23.5          41.8          12.3          42.1          63.55
 8261.79       33580.6


TIME:  0.23 SECS.
```

**Commentary on Program 2-2.** This program was printed out on a terminal printer. Output format varies slightly from one computer facility to another, and thus your printout may not be exactly as shown. The system command, LIS (or LIST), is typed by the programmer whenever one wants a *list*ing of the program in immediate memory. After reading a new program into memory, it is usually wise to list it and obtain a clean copy of the program for the purpose of editing and correcting it.

The source program begins with a READ statement that demands values for $x_1, x_2, \ldots, x_5$. These values are supplied by the DATA statement that follows. The sum of squares, $S_1$ is accumulated in statement 30 and the square of the sum is stored in memory location $S_2$ by statement 40. The symbol ** indicates exponentiation. Many microsystems use ↑2 to square a variable. Statement 50 prints the input data. Statement 60 prints out the solution to the problem, the sum of squares, and the square of the sum in that order. The output is rather sparse and would be impossible to interpret by anyone not knowing how the program is written. Alphanumeric identification would be necessary to identify the entries in the output table for a reader who does not have a listing of the program or who may not read BASIC.

The system command RUN or RUNNH runs the program and causes the PRINT statements to be executed, after which the total run time is given and the statement READY indicates that the computer is ready to go on to a new task. A new data set may be run by simply typing in a new 20 DATA statement. Upon hitting *return*, the new data set will be substituted for the old one and the computer will signify READY. The system command RUN runs the new set. This process can be repeated indefinitely.

## Exercise 2-4

Determine the range, arithmetic mean, median, mode, and quadratic mean of the data set: 7.2, 8.1, 7.5, 9.0, 6.8, 7.2, 7.8.

Solution 2-4

$$\text{Range} = 9.0 - 6.8 = 2.2$$
$$\text{Mean, } \bar{x} = 7.7$$
$$\text{Median} = 7.5$$
$$\text{Mode} = 7.2$$
$$\text{Quadratic mean, } \bar{x}_{rms} = \{1/7\,(51.84 + 65.61 + 56.25$$
$$81.0 + 46.24 + 51.84 + 60.84)\}^{1/2} =$$
$$\{413.63/7\}^{1/2} = 7.7$$

## Program 2-3

```
10   REM PROGRAM TO COMPUTE THE MEAN OF FIVE NUMBERS
20   READ X1,X2,X3,X4,X5
30   DATA 29.56,29.44,29.61,29.51,29.70
40   LET S=X1+X2+X3+X4+X5
50   LET M=S/5.
60   PRINT' THE ARITHMETIC MEAN OF THE FOLLOWING INPUT DATA:'
70   PRINT X1,X2,X3,X4,X5
80   PRINT' IS'
90   PRINT M
100 PRINT' AND THE SUM IS:'
110 PRINT S
120 END

READY
RUNNH

 THE ARITHMETIC MEAN OF THE FOLLOWING INPUT DATA:
 29.56          29.44          29.61          29.51          29.7
 IS
 29.564
 AND THE SUM IS:
 147.82


TIME:   0.23 SECS.
```

Commentary on Program 2-3. This program is similar to Program 1.2 in its first four steps, which compute the sum of the five input data. Statement 50 computes the mean by dividing S by 5, and statements 60 through 110 provide for output of the data set, the mean, and the sum with appropriate alphanumeric identification.

Considerable variation is permissible in the input format. Splitting the data set into parts as in the following variation is acceptable so long as each part has a separate and sequential statement number and each part is preceeded by the word DATA. Splitting up data sets is useful when they are large because it is easier to correct a short DATA statement than it is to correct a long one.

```
30  DATA 29.56,29.44,29.61
31  DATA 29.51,29.70
```

DATA statements may be placed arbitrarily in a program as in the variation below

```
30  DATA 29.56,29.44,29.61
40  LET S=X1+X2+X3+X4+X5
50  LET M=S/5.
51  DATA 29.51,29.70
```

It is often convenient to place the DATA statement at the beginning or just before the END statement. If too many data are input, the computer simply ignores the excess data points.

One must not, however, have fewer data in the DATA statement than are called for by the READ statement. For example,

```
20  READ X1,X2,X3,X4,X5
30  DATA 29.56,29.44,29.61
40 END

READY
RUNNH


? OUT OF DATA IN LINE 20


TIME:  0.14 SECS.
```

does not run. This output indicates that READ statement 20 is the one that is not satisfied by the DATA statement. In our sample program, statement 20 is the only possibility, but in a complicated program with many READ statements and much data, this error statement may be very convenient for tracking down the missing datum or data.

## Program 2-4

Thus far the programs presented have solved problems that were either artificial or very restricted in nature. The following two programs solve a problem frequently encountered in real situations, determination of the mean of $N$ data points, where $N$ may change from one data set to the next, may be very large, or may even be unknown.

From this point on, readers would be well advised to keep a notebook containing the printouts or listings of all programs they have written. The source-program listing can be cut out, pasted into a loose-leaf notebook, and updated as the program is modified. This procedure not only provides an index to the programs already written, but provides ready made modules to insert into new programs. Certain tactical problems, such as reading in data sets and writing out results, occur time and time again in programming; hence, it is common procedure to copy entire modules from old programs into new ones. If facilities exist for doing so, a permanent library should be kept on disc or tape for future use.

The problem of finding the mean of *N* numbers is solved below for two conditions: *N* known and *N* unknown. Since both programs solve the same problem, the strategy must be the same for both; and since both are in BASIC, the language is also the same. In these two programs, we are concentrating our attention on the tactics of programming, which, it will be seen, differ considerably between them.

```
5 DATA 3,1,2,4
10 REM PROGRAM TO CALCULATE THE MEAN OF N NUMBERS, N IS KNOWN
20 READ N
30 S=0.0
40 LET I=0
50 READ X
60 LET S=S+X
70 LET I=I+1
80 IF I<N, THEN 50
90 LET M=S/N
100 PRINT M
110 END

READY
runnh


 2.33333


TIME:  0.11 SECS.
```

**Commentary on Program 2-4.** To make substitutions easy, the DATA statement has been placed before the body of the program. This DATA statement supplies *two* READ statements, the integer 3 is *N*, and the real variables 1, 2, and 3 constitute the data set.

This program is quite distinct from the previous ones in that it is open-ended. In principle, there is no limit to the number of data that can be treated by this small program, although problems of punching in the data and computer memory limitations would set some kind of practical limits. A distinct advantage to batch processing is the ease with which incorrectly punched cards can be pulled out of a data deck and replaced. Frequently in research, very large data decks are accumulated and revised over many months or even years. The same kind of accumulation of data using forms of memory other than punched cards, e.g., disc memory, is also possible.

The first READ statement demands the reading of *N*, the number of points in the data set. The value of *N* has been set at 3 for simplicity's sake in this case, although it may be large. Because READ $N$ comes first in the program, the first number in the data set must be $N$. After that, order is not important in this program using ungrouped data. The quantity $S$ is initialized at zero, as is $I$. Note that statements 30 and 40 differ in the use or the absence of the word LET. Both statements store zero in memory, one in location $S$ and one in location $I$. The form of statement 30 is not valid in some forms of BASIC, but statement 40 is universal.

Statement 50 reads the first value of $x$ and statement 60 adds it to the contents of location $S$ (zero). After statement 60 has been executed, location $S$ contains 1. Statement 70 increments $I$ by 1. Now both locations contain 1. The key decision in this program is made at statement 80. If $I$ is less than $N$, then control returns to statement 50 and another value of $x$ is read. On the second iteration, $x_2$ is added to the contents of $S$, and on the third, $x_3$ is added to $S$, resulting in the accumulation of the sum of $x_1$, $x_2$, and $x_3$ in $S$. On the third iteration, however, $I = N$, and control is not looped back to statement 50. Instead the computer continues to execute commands in the order of their appearance in the program. Control goes to statement 90, which computes the mean, followed by statement 100, which prints it. In some forms of BASIC, the comma is deleted from statement 80.

## Program 2-5

($N$ unknown) Program 2-5 for the arithmetic mean must be written using a new set of tactics because $N$ is unknown. No scheme analogous to that used in Program 2-4 can be used because we do not know the number of cards in the data deck or the number of data stored on a disc. If the deck is large enough, simply counting by hand may be impractical. A separate program could be written to count the data deck, but the method used in Program 2-5 avoids the necessity of two separate runs and computes the mean and the count at the same time. The data set must be such that no one of its members is 99999. This device for exiting from a read loop is most useful in batch processing a large deck of cards for which $N$ is unknown. One simply places a 99999 "trailer card" at the end of the deck to achieve a proper exit from the loop.

```
COMP2.3          12:58          18-DEC-79


10   REM PROGRAM TO CALCULATE THE MEAN OF N NUMBERS, N UNKNOWN
20   S=0.0
30   N=0
40   READ X
50   DATA 1,2,4,99999
60   IF X=99999, THEN 100
70   S=S+X
80   N=N+1
```

```
90  GO TO 40
100 M=S/N
110 PRINT M,N
120 END

READY
runnh

 2.33333       3


TIME:  0.13 SECS.
```

**Commentary on Program 2-5.** Statements 20 and 30 initialize $S$ and $N$, and statement 40 reads the first value of $x$ from the data set given in statement 50. The key decision-making step in Program 2-5 is statement 60, which tests the input datum to see whether it is equal to 99999. If not, the program steps are executed in the order in which they occur. Statement 70 places the first datum in location $S$, and 80 increments $N$ by 1, thereby counting the first datum. Statement 90 is an *unconditional* GO TO that sends control back to statement 40 to read another datum. The second datum is added to $S$ and counted. This process is continued indefinitely until the datum 99999 is found. Control is then *branched* to statement 100, effectively jumping over the accumulator step (70), the counting step (80), and the GO TO step. Clearly, if 99999 is or may be in the data set, another test condition must be found.

To understand how control is being directed in this program, it is necessary to remember that control passes from one step to another according to their numerical sequence *unless directed to do otherwise*. The GO TO 40 statement is unconditional; it always directs control to statement 40, hence it must be jumped to exit from the loop. The IF. . .THEN statement is conditional; it directs control to jump to statement 100 if a certain arithmetic condition is satisfied ($x_i = 99999$), otherwise, the normal flow of the program is uninterrupted. Statement 100 computes the mean and 110 prints it, as we have previously seen. For reasons of space and simplicity, the output in many programs in this book will not be documented with alphanumeric headings and explanations. Those that are so documented should serve as models for the rest. In any case, documentation is a matter of preference and will differ from one programmer to another.

## The Mean of Means

It is not uncommon for us to want to determine the mean of a data set that is comprised of means of other data sets. A common example arises when we wish to take the mean of experimental determinations of a quantity that has been measured by several different research groups. Each group reports its mean value taken from several separate determinations of the quantity of interest, and we wish to determine the grand mean.

Suppose we work the problem for two means and generalize to many means. Data set one, containing $N_1$ members, has a mean $\bar{x}_1$ and data set two, containing $N_2$ members, has $\bar{x}_2$. The mean of all the data is

$$\overline{x_{\text{all}}} = \frac{(x_{a1} + x_{a\,2} + \ldots + x_{aN1}) + (x_{b1} + x_{b2} + \ldots + x_{bN2})}{N_1 + N_2} \quad (2\text{-}18)$$

The two individual means are

$$\overline{x_1} = \sum_{i=1}^{i=N_i} x_{ai} \Big/ N_1 \quad (2\text{-}19)$$

and

$$\overline{x_2} = \sum_{j=1}^{j=N_2} x_{b_j} \Big/ N_2 \quad (2\text{-}20)$$

or

$$\sum_{i=1}^{i=N_1} x_{a_i} = N_1 \overline{x_1} \quad (2\text{-}21)$$

and

$$\sum_{j=1}^{j=N_2} x_{b_j} = N_2 \overline{x_2} \quad (2\text{-}22)$$

but the summations of $x_{ai}$ and $x_{bi}$ are precisely the summations we need to add together to obtain the numerator in Eq. (2-18) for $x_{\text{all}}$

$$\overline{x_{\text{all}}} = (N_1 \overline{x_1} + N_2\, \overline{x_2}) \Big/ (N_1 + N_2) \quad (2\text{-}23)$$

In general

$$\overline{x_{\text{all}}} = \sum N_k\, \overline{x_k} \Big/ \sum N_k \quad (2\text{-}24)$$

where $k$ is the running index from one to the number of means to be included in the grand mean.

The procedure given above essentially weights each individual mean according to the number of data points contributing to it. Frequently (see Problems), the number of data points contributing to all individual means is the same. The grand mean of means can be computed simply as

$$\overline{x_{\text{all}}} = \sum \overline{x_k} \Big/ M$$

where $M$ is the number of individual means included in the grand mean.

## Program 2-6

In a previous example, we had reason to determine the range of a data set by subtracting the smallest member of the set from the largest. The smallest and largest members of a small set may be selected by inspection, but the same task for a sufficiently large set may require a machine. One method of approach is illustrated in Program 2-6 and a modification thereof.

```
1    REM PROGRAM TO FIND THE LEAST OF FIVE NUMBERS
10   READ A,B,C,D,E
20   DATA 2,3,4,5,6
30   LET L=A
40   IF B>L, THEN 60
50   LET L=B
60   IF C>L, THEN 80
70   LET L=C
80   IF D>L, THEN 100
90   LET L=D
100  IF E>L, THEN 120
110  LET L=E
120  PRINT L
130  END

READY
RUNNH


 2


TIME:   0.08 SECS.
```

Commentary on Program 2-6. After labeling the program, reading five input data, and placing them in five separate memory locations, A, B, C, D, and E, step 30 retrieves the datum in location A and copies it into memory location L. Step 40 retrieves the datum in location B and tests it to determine whether it is larger than the datum in location L. There are two possibilities (1) B is larger than L, whereupon the program skips to step 60, leaving datum A in location L. (2) B is not larger than L, whereupon the normal sequence of steps is followed. The next sequential step, step 50, substitutes B for A in location L. Datum A is lost from memory location L, where it is replaced by B, and does not appear at that location again during the running of the program.

Steps 60 and 70 perform the same test on the datum in location C. If C is less than or equal to the datum in L, C is copied into location L. If it is greater than L, the program skips ahead to step 80 and performs the same test on the datum in D. The datum in E is tested the same way, completing the testing of all five data. At this point, the data have been surveyed and the smallest datum of the set is in location L. Although they are no longer needed, the original data are still stored sequentially in locations A through E. It is important to remember what is at each location because, in later programs, we may have cause to refer back to or operate on the original data several times in the running of the program. Step 120 prints out the answer and 130 ends the program.

## Program 2-6A

```
1    REM PROGRAM TO FIND THE RANGE OF A FIVE NUMBER DATA SET
10   READ A,B,C,D,E
15   DATA 8,3,1,0,5
20   LET L=A
30   IF B>L, THEN 50
40   LET L=B
50   IF C>L, THEN 70
60   LET L=C
70   IF D>L, THEN 90
80   LET L=D
90   IF E>L, THEN 110
100 LET L=E
110 LET M=A
120 IF B<M, THEN 140
130 LET M=B
140 IF C<M, THEN 160
150 LET M=C
160 IF D<M, THEN 180
170 LET M=D
180 IF E<M, THEN 200
190 LET M=E
200 PRINT "THE RANGE COVERED  BY THE DATA SET IS"
210 PRINT L, M
220 END

READY
runnh


THE RANGE COVERED  BY THE DATA SET IS
 0                8


TIME:  0.21 SECS.
```

**Commentary on Program 2-6A.** The modification of program 2-6A consists of two modules, one that selects the lowest value of a data set, and one that selects the largest datum of the set. The first module was lifted directly from Program 2-6 and the second module is essentially the same as the first except that the sign $<$ has been substituted for $>$. This scheme is not self-terminating and can be extended merely by adding F, G, . . . to the READ and DATA statements and adding testing modules like steps 30 and 40 to the program between the LET L $=$ E and PRINT L statements. Although there are better ways of doing it, this program logic can be extended to select the least (or greatest) of any number of data up to the limit imposed either by the number of variable names permitted in the language used or by the number of computer memory locations available. Once the least and the greatest of a data set have been selected, the range is known.

## Exercise 2-5

A group of 18 hospital patients received intravenous injections of glucose in units of grams per kilogram of body weight per hour. Their retention of

Table 2-1
Injection Rate and Retention of Glucose for a Group of Eighteen Hospital Patients

| Patient | Injection rate, g/kg-h | Retention, g/kg |
|---|---|---|
| 1 | .071 | 0.070 |
| 2 | .155 | 0.150 |
| 3 | .217 | 0.212 |
| 4 | .382 | 0.284 |
| 5 | .453 | 0.448 |
| 6 | .501 | 0.490 |
| 7 | .737 | 0.675 |
| 8 | .907 | 0.815 |
| 9 | 1.107 | 0.803 |
| 10 | 1.136 | 1.019 |
| 11 | 1.169 | 0.853 |
| 12 | 1.274 | 1.043 |
| 13 | 1.296 | 1.109 |
| 14 | 1.430 | 1.401 |
| 15 | 1.558 | 1.411 |
| 16 | 1.787 | 1.280 |
| 17 | 1.920 | 1.913 |
| 18 | 2.171 | 1.533 |

glucose was monitored and the results are shown in units of grams retained per kilogram of body weight. Devise a computer program that will determine the arithmetic mean rate of glucose injection and the arithmetic mean retention for the group. Note that although the data for injection rate are ordered from the lowest to the highest in a regular way, the retention shows irregularities. This occurs because the values for injection rate and retention are not perfectly correlated.

## Glossary

*Accuracy*. Agreement of an experimental result with the true result.

*Arithmetic Mean*. Sum of a set of measurements divided by the number of members in the set.

*Array* (statistician's meaning). Data set presented according to some specific pattern, usually from the smallest member of the set to the largest; (computer programmer's variant): Any fixed sequence, usually not from the smallest to the largest.

*Continuous Variable*. Variable capable of taking on any of an infinite number of values over some interval, as distinguished from a discrete variable (mathematically, if $F(x)$ approaches $F(A)$ as $x$ approaches $A$, $F(x)$ is continuous at $A$).

*Derivative.* Rate of change of the function $F(x)$ with an infinitesimal change of $x$.

*Discrete Variable.* Variable that can take on only a finite number of values over an interval, as distinguished from a continuous variable.

*Euler's Theorem.* Equation giving the total variation of a function of many variables in terms of each taken individually, holding all others constant.

*Grouped Data.* Data presented in quantitatively distinguishable subsets, usually arbitrarily defined.

*Infinitesimal.* An immeasurably small change in a continuous variable.

*Insignificant Figures.* Figures that imply an accuracy beyond the accuracy of the least accurate measurement from which they are derived.

*Integer Variable.* A variable that is restricted to the integers 0, 1, 2, 3,. . . or, less commonly, the negatives of these.

*Interval.* A subdivision of the range covered by a variable.

*Median.* The point in an array that divides it into two equal parts.

*Mode.* The member of a set that appears most frequently.

*Partial Derivative.* Rate of change of the function $F(x_1, x_2, \ldots x_N)$ with an infinitesimal change in one of the independent variables, $x_i$, holding all others constant.

*Partial Differentiation.* The mathematical operation of taking a partial derivative.

*Precision.* Agreement of a series of experimental results with each other as contrasted to accuracy (see above).

*Quadratic Mean.* The square root of the mean of the squares of a data set.

*Random.* Of a process of selection in which each member of a set has an equal probability of being chosen; not biased.

*Random Errors.* Errors of indeterminate source that are not biased in magnitude or direction.

*Range.* The difference between the largest and the smallest members of a set.

*Real Variable.* A continuous variable belonging to the real number set; not complex.

*Root Mean Square.* Quadratic mean.

*Running Index.* Counter used to differentiate members of a set; in the set $x_1, x_2, \ldots, x_i, \ldots x_N$, $i$ is the running index.

*Significant Figures.* Digits that reflect the accuracy of a measurement.

*Systematic Errors.* Errors, generally arising in the measuring device, that are biased in magnitude and direction.

*Ungrouped Data.* Data presented in random order.

*Variation.* Change in a variable.

*Weighted Arithmetic Mean.* Mean in which some members of the set are multiplied by weighting factors so as to make them have a larger or smaller effect on the calculated result.

## Problems

*1*. Add the following numbers, retaining the proper number of significant figures.

(a) 23.97 + 4.201
(b) 6.02E24 + 5.9E23
(c) 0.0913 + 1.83 + 7.11E−3

*2*. Multiply the following numbers, retaining the proper number of significant figures.

(a) $0.340 \times 27.9 \times 1.3 \times 10^{-4}$
(b) $7.2\text{E}4 \times 7.2 \times 10^{5}$
(c) $8.34 \times 2.01$

*3*. Round the following experimental readings to four significant figures: 40.31, 100.49, 1364.2, 50955.

*4*. The area of a triangle is given by $A = 0.5 \times B \times H$, where $B$ is the base of the triangle and $H$ is its height. Derive an equation for the fractional error in $A$ with simultaneous errors in $B$ and $H$; this situation is analogous to that treated by Eq. (2-5).

*5*. Write down Euler's Theorem for the variation in $A$ with simultaneous variations in $B$ and $H$ for the area of a triangle, $A = 0.5 \times B \times H$.

*6*. Determine the range, arithmetic mean, median, mode, and quadratic mean of the data set 1.448, 1.456, 1.399, 1.412, 1.424, 1.441.

*7*. Write a program in BASIC to compute the sum of the squares of the numbers 5.0, 5.5, 5.2, 5.6, 6.0.

*8*. Suppose a cylinder of radius 1.000 m and height 1.000 m is measured, but that the measured $r = h = 1.001$ m. Calculate the error in volume and the % error in volume by Eq. (2-3) and by its approximate form, Eq. (2-4). Is Eq. (2-4) a good approximation to Eq. (2-3) for this case? Since the errors are known exactly (by hypothesis), calculate your answer to four significant figures.

*9*. "Document" Program 2-2 by including REM statements to explain the function of statements 30 and 40 in the listing and PRINT statements to make the output clearer.

*10*. Rearrange and modify Program 2-5 so that a provisional mean is printed out for each input datum, i.e., one that generates the means 1, 1.5, and 2.33333 for the data set 1. 2. 3.

*11*. Write a program to compute and print out $x$, $x^{1/2}$, and $(x^{1/2})^2$ for x = 1 to 1000. Comment upon truncation errors encountered if any.

## Bibliography

R. L. Albrecht, L. P. Finkel, and J. R. Brown, *BASIC*, Wiley, New York, N.Y., 1973.

J. R. Barrante, *Applied Mathematics for Physical Chemistry,* Prentice-Hall, Englewood Cliffs, N.J., 1974.

J. Hennenfield, *Using BASIC,* Prindle, Webber & Schmidt, Boston, Mass., 1978.

D. A. Skoog and D. M. West, *Analytical Chemistry,* 3rd ed, Saunders, Philadelphia, Pa., 1980.

C. E. Swartz, *Used Math for the First Two Years of College Science,* Prentice-Hall, Englewood Cliffs, N.J., 1973.

H. D. Young, *Statistical Treatment of Experimental Data,* McGraw-Hill, New York, N.Y., 1962.

# Chapter 3

# Understanding Experimental Data Dispersion

Few, if any, scientific conclusions rest on single experimental measurements, hence some method is needed to express the dispersion of experimental results about some measure of central tendency, typically, the arithmetic mean. The range is one possibility. If the arithmetic mean of a large set of experimental data is 123 with a range of 110–135, we would assume, and usually rightly, that the *scatter* or experimental error of this set is less than that of a set with $\bar{x} = 123$ and a range of 100–150.

Though the range is sometimes used in informal conversation, it is not particularly useful as a rigorous measure of dispersion of a data set because one or two extreme variations can give a wide range to a set of data that are otherwise closely grouped. It is possible, though perhaps not likely, that the condition reached in the previous paragraph, using the range as a measure of dispersion, is wrong. Suppose the first set had data scattered evenly over the range 110–135, whereas the second (large) data set had many values very close to 123 and two extreme values, one of 100 and one of 150. The range has not correctly reflected the relative scatter of these two sets and has led us to a conclusion opposite to the correct one.

## The Average Deviation

The error of any single measurement is its *deviation, d,* from the true result, $x_t$. Usually, $x_t$ is not known. We suppose, however, that the arithmetic mean approaches $x_t$ for a large data set on the basis, merely, of cancellation of random errors, some high and some low. If we replace the hypothetical or inaccessible $x_t$ with $\bar{x}$, which is readily calculated for any data set, we may write the *deviations from the mean* for members of the set

$$d_i = x_i - \bar{x} \tag{3-1}$$

As $N \rightarrow \infty$, $\overline{x} \rightarrow x_t$ and a representative or "average" $d_i$ approaches the experimental error, provided that errors are random. The question is what kind of "average" to take.

The arithmetic mean deviation, $\bar{d}$ (frequently called the *average deviation*) is

$$\bar{d} = \sum d_i / N = \sum (x_i - \bar{x}) / N \tag{3-2}$$

It is readily proven that the arithmetic mean deviation is zero (see problems), hence it can hardly serve as a measure of dispersion about the arithmetic mean.

### Exercise 3-1

Write a program in BASIC to obtain the sum of the deviations from the mean of a set of 12 data.

## The Mean Deviation Independent of the Sign

One way of getting around cancellation of deviations with opposite signs is simply to drop the sign. The magnitude of any number considered without regard to its sign is its *absolute value* and is denoted by vertical lines on either side of the number.
Thus

$$|\bar{d}| = \sum |d_i| / N = \sum |x_i - \bar{x}| / N \tag{3-3}$$

is the *arithmetic mean absolute deviation*. It does not become zero for any nonidentical data set, it is large when experimental scatter is large and small when experimental scatter is small, hence it is valid as a measure of dispersion.

The arithmetic mean absolute deviation, frequently simply called the *mean deviation*, is discussed in many elementary laboratory manuals in chemistry and biology, but it does not strictly apply to random errors of a continuous variable. It has been widely and erroneously applied to randomly distributed experimental data, with the result that some journals no longer accept it as a measure of data dispersion. This is allowing the pendulum to swing too far in the reverse direction because there are some distributions for which the arithmetic mean absolute deviation is the appropriate measure of dispersion.

### Exercise 3-2

Write a problem in BASIC to determine the mean deviation independent of the sign of 12 data. Use the same data set you used in Exercise 3-1.

## Population Parameters and Sample Statistics

There are two distinct situations in data gathering that will concern us in this book. Tests may be made and data recorded for all members of a set or class of individuals or they may be recorded for a subset containing fewer members than the entire set. The entire set of individuals possessing a common characteristic that differentiates them from other sets, but which they possess in common, is frequently called a *population* or a *universe*. When not all members of a set are considered, those members that are considered are collectively called a *sample*. Samples are usually taken to represent the entire population although this is not always true. Samples that do not represent the population from which they were drawn are said to be *biased*. Considerable care should be taken to avoid biased samples or to take account, mathematically, for their bias.

Most samples the physical scientist or life scientist encounters are finite, but so large that it is not practical to obtain data on the entire population. Thus, the arithmetic mean calcium content of the blood of a finite sample of healthy individuals is taken to represent the arithmetic mean calcium content of the blood of all healthy individuals, a population that is finite but very large. Population means are given the special symbol $\mu = \Sigma x_i/N$ where $N$ is the number of members in the *entire* population. If the distribution of data is random about the sample arithmetic mean $\bar{x}$, it approaches the population arithmetic mean $\mu$ as $N$ becomes very large.

## The Standard Deviation

When deviations about a measure of central tendency are random, the appropriate measure of dispersion is the quadratic mean of the deviations, called the *standard deviation*

$$\sigma = \left[\sum d_i^2 \Big/ N\right]^{1/2} = \left[\sum (x_i - \mu)^2 \Big/ N\right]^{1/2} \tag{3-4}$$

for a population or

$$s = \left[\sum d_i^2 \Big/ (N - 1)\right]^{1/2} = \left[\sum (x_i - \bar{x})^2 \Big/ (N - 1)\right]^{1/2} \tag{3-5}$$

for a sample.

Squaring both sides of Equations (3-4) or (3-5), we obtain a measure of dispersion called the *variance*

$$\sigma^2 = \sum d_i^2 \Big/ N \tag{3-6}$$

or

$$s^2 = \sum d_i^2 \Big/ (N - 1) \tag{3-7}$$

The variance lends itself more readily to some algebraic and computational methods and, of course, whenever we have either the variance or the standard deviation, we can easily obtain the other.

Some types of machine calculations (particularly those employing programmable calculators or microcomputers with limited memory) are facilitated by writing the variance in a different but equivalent way. From Eq. (3-6), we have

$$\sigma^2 = (1/N) \sum (x_i - \mu)^2 = (1/N) \sum (x_i^2 - 2x_i \mu + \mu^2) \quad (3\text{-}8)$$

The sum can be broken up into a three-term equation, each term being, itself, a sum

$$\sigma^2 = (1/N) \sum x_i^2 - (1/N) \sum 2x_i \mu + (1/N) \sum \mu^2 \quad (3\text{-}9)$$

We note that $\mu$ is the same in each term of the middle sum; hence it can be factored out. Also, $\mu^2$ is the same in each term of the rightmost sum; hence

$$(1/N) \sum \mu^2 = (1/N)N\mu^2 \quad (3\text{-}10)$$

which leads to

$$\sigma^2 = (1/N) \sum x_i^2 - \mu^2 \quad (3\text{-}11)$$

This form is necessary when one is working under computer memory limitations because as data are fed into the computer it is only necessary to store $N$, $\Sigma x_i^2$, $\Sigma x_i$, and $\mu^2$ to calculate the value of $\sigma^2$. Equation (3-5) requires storage of all input data because the values of $d_i$ are not computed until a final value of $\mu$ has been obtained. For sets of data that exceed the number of memory locations a machine has, this is not possible.

## Exercise 3-3

Show that Eq. (3-11) follows from Eq. (3-9).

## Program 3-1

Because the standard deviation is calculated for almost every numerical data set reported in the scientific literature, we shall include a program to calculate it here. There are many tactics by which $s$ may be calculated. The following program shows one method, using the INPUT statement, a statement which is useful for *interactive* programs, i.e., programs in which the computer user engages in a "dialog" with the computer.

```
10  REM MEAN AND STANDARD DEVIATION USING THE INPUT STATEMENT
20  S=0.0
25  N=0
30  INPUT X
40  S=S+X
50  N=N+1
60  M=S/N
70  D=X-M
80  D1=D1+D
90  S1=SQRT(D1**2/(N-1))
100 PRINT M;S1
110 GO TO 30
200 END

READY
runnh

?10.23

% DIVISION BY ZERO IN LINE 90
 10.23   1.30438E+19
 ?10.12
 10.175   5.50001E-2
 ?10.46
 10.27   9.54593E-2
 ?10.39
 10.3   0.129904
 ?10.55
 10.35   0.2125
 ?10.42
 10.3617   0.216153
 ?
```

Commentary on Program 3-1. The initialization statements 20 and 25 are not necessary in many systems because referenced memory locations are automatically initialized to zero when the RUN command is entered. This feature is not universal, particularly on smaller machines, and until it is, it is good form to include initialization statements. If a divide by zero error is encountered, initialize $N$ at a very small nonzero value, e.g., 0.000001. After initialization, we encounter the INPUT statement. When the program is executed, control halts at this statement and a question mark appears on the CRT or is printed out by the terminal printer. When a datum is typed in and the programmer hits the return key, that datum is taken as $X$ and control passes on to the succeeding steps in the usual way. Statements 40 and 50 accumulate the first value of $X$ in $S$, 1 in $N$ and 60 computes the arithmetic mean ($X/1$ on the first iteration). Statements 70 and 80 accumulate the deviations, 90 computes the standard deviation for as many data of the sample as have been entered on each iteration. Some systems use SQR to compute the square root.

These three steps are not meaningful on the first iteration because the standard deviation is not meaningful for one datum. Note the error message on the first input. To see how the program functions meaningfully, we must consider the third or higher iteration. On the third iteration, for example, the programmer responds to the input prompt ? with the third input datum, 10.46. That datum is added to $S$, which already contains the sum of $X(1)$ and $X(2)$ giving the sum of three data. Now $N = 3$ and both $M$ and $S1$ are meaningful numbers, 10.27 and 0.10 (rounded). After the PRINT statement, 110 sends control back to 30 INPUT $X$ for another datum and a new mean and standard deviation are computed and printed. This dialog goes on indefinitely; the computer asks for a new datum, the operator supplies it, the computer replies with a new $\bar{x}$ and $s$ and so on. As the program is written, it constitutes an infinite loop, considered very bad programming form. To terminate this program, one must abort it. This is done in some systems by pressing down the control button on the keyboard and hitting the letter C twice. Other systems have a BREAK key. One of the problems at the end of the chapter asks you to provide an exit from the INPUT loop so that the program does not have to be aborted to be terminated.

## Relative Standard Deviation

The fractional standard deviation, $s/\bar{x}$ and the percent standard deviation, called the *coefficient of variation*,

$$\%s = (s/\bar{x})100 \tag{3-12}$$

are frequently used. These unitless ways of expressing dispersion facilitate comparison of experimental methods. Thus, if we are told that two methods of chemical analysis have a standard deviation of 0.1 g, we must know how many grams of material were analyzed to compare the methods. A fractional or percent standard deviation already contains that information so that we can say that two methods having percent $s$ of 0.1 are equivalent. Comparison by percent $s$ is not valid, however, for samples that vary widely in size. A 1% standard deviation may be unacceptable for a method used to determine one gram of a sample, but it may be very good for a method that is used to determine $10^{-3}$ or $10^{-4}$ grams.

### Exercise 3-4

Write a program that prints out the standard deviation and the percent standard deviation.

## Grouped Data and Frequency Distributions

Unless it is quite small, a table of raw data as it has been collected in the laboratory or in the field is not very informative. Arranging the data into an *array* from lowest to highest value makes it more informative. The range is

Table 3-1
Number of Children in Ten
Arbitrarily Selected Families

| Family | Number of children |
|---|---|
| A | 2 |
| B | 1 |
| C | 0 |
| D | 3 |
| E | 2 |
| F | 2 |
| G | 0 |
| H | 1 |
| I | 7 |
| J | 4 |

evident from an array, and a cluster or clusters of data points are easy to spot. Presentation of data in this form is wasteful of space, however, hence tabular data presentation is often in the form of frequency groups. In grouping data, we select a set of intervals covering the entire range and report the frequency with which data points appear in each interval. Usually, the intervals are of equal size. In simple cases, no information is lost by this procedure and the data are presented in a clearer and more concise way than they were in the form of a raw data table. To illustrate, suppose we investigate the number of children in 10 families and our raw data findings are as shown in Table 3-1.

In the form of a frequency table, we have Table 3-2.

Table 3-2
Frequency Distribution of the Number
of Children in Ten Arbitrarily Selected
Families

| Number of children | Frequency, $f$ |
|---|---|
| 0 | 2 |
| 1 | 2 |
| 2 | 3 |
| 3 | 1 |
| 4 | 1 |
| 5 | 0 |
| 6 | 0 |
| 7 | 1 |
| | 10 |

Like all illustrations used in the text of this book, the data set in Tables 3-1 and 3-2 was kept small for the sake of simplicity. For this data set, the advantage of using a frequency distribution is not great because the number of ages represented is not much smaller than the number of families represented. If the reader will imagine data sets containing thousands or even hundreds of thousands of data points such as might be produced by a census report, the advantage of the frequency distribution becomes clear. Frequency distributions are useful in situations in which there is a limited number of frequency groups, but in which the number of input data may be indefinitely large.

## Program 3-2

Transforming raw data sets into frequency distributions is an important computer application. Program 3-2 introduces a method of data storage using a DIM (dimension) statement and a FOR. . .NEXT read loop. There are significant advantages to this method of data input.

```
1     REM DIMENSIONS AND ARRAYS: PRELIMINARY PROGRAM
5     DATA 4,5,6,3,4
10    DIM X(100)
20    FOR I=1 TO 5
30    READ X(I)
40    NEXT I
50    FOR I=1 TO 5
60    PRINT X(I)
70    NEXT I
80    END

READY
runnh


 4
 5
 6
 3
 4



TIME:   0.12 SECS.
```

**Commentary on Program 3-2.** After a REM and a DATA statement, 10 DIM X(100) establishes the dimension of the subscripted variable $X$ at 100. This dimension statement reserves 100 locations in computer memory for $x_1, x_2, \ldots x_{100}$. Unless there is ambiguity, the first three letters of a statement, DIM in this case, often suffice for the whole word. If no dimension statement is present, the dimension is taken arbitrarily to be 15 in many systems. Taking an arbitrary value for anything not specifically stated in a program is called establishing that value *by default*. Statements 20 through 40 constitute a FOR. . .NEXT read loop. Comparison with FOR. . .NEXT loops we have already studied should make it evident that on the first iteration, I = 1 and the first number after 5 DATA is stored in location $X(1)$. On the second, a datum is stored in $X(2)$ and so on until all five data have been stored. The dimension statement must be equal to or larger than the number of data. One usually gives a dimension much larger than necessary, preferring to waste some memory locations rather than to have the data of some future set overflow the dimension, resulting in an error statement and no run.

Statements 50–70 constitute a FOR. . .NEXT print loop that functions just as the read loop did except, of course, that $X(1)$. . .$X(5)$ are printed out in the order in which they were stored. Because there is a specific order to the input data which is retained in the output, the data set $X(I)$ is an *array*.

## Program 3-3

Program 3-3 uses the input tactic of Program 3-2 as part of a program that takes the 20 integer data points in Table 3-3 and prints out the frequency of data points at each integral age: 0, 1, 2,. . .up to age 50, an interval considerably larger than the range. From the computer output, one can construct tables like Table 3-4.

Table 3-3
Ages of Parents in Ten Arbitrarily Selected Families

| | Age | |
|---|---|---|
| Family | Male parent | Female parent |
| A | 25 | 23 |
| B | 42 | 38 |
| C | 35 | 36 |
| D | 41 | 41 |
| E | 21 | 19 |
| F | 39 | 34 |
| G | 49 | 42 |
| H | 34 | 33 |
| I | 32 | 33 |
| J | 28 | 26 |

Table 3-4
Frequency Distribution of the Ages of Parents in Ten Arbitrarily Selected Families Without Distinction According to Sex

| Age group | Frequency |
|---|---|
| 16–20 | 1 |
| 21–25 | 3 |
| 26–30 | 2 |
| 31–35 | 6 |
| 36–40 | 3 |
| 41–45 | 4 |
| 46–50 | 1 |

```
1    DIM X(100)
5    DATA 25,42,35,41,21,39,49,34,32,28,23,38,36,41,19,34,42,33,33,26
10   REM FREQUENCY DISTRIBUTION
20   T=0
30   FOR I=1 TO 20
40   READ X(I)
50   NEXT I
60   FOR J=1 TO 50
70   F=0
80   T=T+1
90   FOR I=1 TO 20
100 IF T=X(I), THEN 110
105 GO TO 115
110 F=F+1
115 NEXT I
120 PRINT 'AT THE VALUE',T,'THE FREQUENCY IS',F
130 NEXT J
140 END

READY
runnh

AT THE VALUE    1          THE FREQUENCY IS          0
AT THE VALUE    2          THE FREQUENCY IS          0
AT THE VALUE    3          THE FREQUENCY IS          0
AT THE .VALUE   4          THE FREQUENCY IS          0
AT THE VALUE    5          THE FREQUENCY IS          0
AT THE VALUE    6          THE FREQUENCY IS          0
AT THE VALUE    7          THE FREQUENCY IS          0
AT THE VALUE    8          THE FREQUENCY IS          0
AT THE VALUE    9          THE FREQUENCY IS          0
AT THE VALUE    10         THE FREQUENCY IS          0
AT THE VALUE    11         THE FREQUENCY IS          0
AT THE VALUE    12         THE FREQUENCY IS          0
AT THE VALUE    13         THE FREQUENCY IS          0
AT THE VALUE    14         THE FREQUENCY IS          0
AT THE VALUE    15         THE FREQUENCY IS          0
AT THE VALUE    16         THE FREQUENCY IS          0
AT THE VALUE    17         THE FREQUENCY IS          0
AT THE VALUE    18         THE FREQUENCY IS          0
```

```
AT THE VALUE   19          THE FREQUENCY IS          1
AT THE VALUE   20          THE FREQUENCY IS          0
AT THE VALUE   21          THE FREQUENCY IS          1
AT THE VALUE   22          THE FREQUENCY IS          0
AT THE VALUE   23          THE FREQUENCY IS          1
AT THE VALUE   24          THE FREQUENCY IS          0
AT THE VALUE   25          THE FREQUENCY IS          1
AT THE VALUE   26          THE FREQUENCY IS          1
AT THE VALUE   27          THE FREQUENCY IS          0
AT THE VALUE   28          THE FREQUENCY IS          1
AT THE VALUE   29          THE FREQUENCY IS          0
AT THE VALUE   30          THE FREQUENCY IS          0
AT THE VALUE   31          THE FREQUENCY IS          0
AT THE VALUE   32          THE FREQUENCY IS          1
AT THE VALUE   33          THE FREQUENCY IS          2
AT THE VALUE   34          THE FREQUENCY IS          2
AT THE VALUE   35          THE FREQUENCY IS          1
AT THE VALUE   36          THE FREQUENCY IS          1
AT THE VALUE   37          THE FREQUENCY IS          0
AT THE VALUE   38          THE FREQUENCY IS          1
AT THE VALUE   39          THE FREQUENCY IS          1
AT THE VALUE   40          THE FREQUENCY IS          0
AT THE VALUE   41          THE FREQUENCY IS          2
AT THE VALUE   42          THE FREQUENCY IS          2
AT THE VALUE   43          THE FREQUENCY IS          0
AT THE VALUE   44          THE FREQUENCY IS          0
AT THE VALUE   45          THE FREQUENCY IS          0
AT THE VALUE   46          THE FREQUENCY IS          0
AT THE VALUE   47          THE FREQUENCY IS          0
AT THE VALUE   48          THE FREQUENCY IS          0
AT THE VALUE   49          THE FREQUENCY IS          1
AT THE VALUE   50          THE FREQUENCY IS          0
```

**Commentary on Program 3-3.** Program 3-3 has the data from Table 3-3 as its input. The statements up to 50 are: dimension, data, remark, initialization and a FOR. . .NEXT read loop which should be familiar. At 60, control enters a loop of 50 iterations which initializes $F$ at zero and increments $T$ by one on each iteration. At 90, control enters a second loop within the first loop. The second loop or *inner* loop is said to be *nested* within the *outer* loop. The running index for the inner loop goes from $I = 1$ to 20, hence we shall call the inner loop the $I$ loop in contrast to the outer loop, which is the $J$ loop.

Concentrate on what the $I$ loop is doing. Since $T$ was incremented to 1 before control entered the $I$ loop, the $I$ loop is essentially a search module of the program that scans the entire data set to see if one or more datum is equal to 1. It does this *via* the 100 IF statement that directs control to 110 $F = F + 1$ if and only if $X(I) = 1$. For no datum is this true, hence 105 GO TO 115 causes control to jump over 110 for each iteration of the $I$ loop and $F$ remains at zero through all 20 iterations. Eventually the $I$ loop is satisfied and control proceeds along the normal route to print the contents of location $F$, i.e., zero. After printing out the result that no parents in the group were 1 year old, control proceeds to 130 but the outer loop is not satisfied so the entire process starts again.

Each time the $J$ loop is iterated, the test age $T$ is increased by one year. For a time, $F$ is always found to be zero, but at $T = 19$, a datum, $X(15) = 19$, is equal to $T$. Statement 110 is activated and $F$, the frequency of parental ages equal to 19, is incremented to 1. As there are no other parental ages of 19, $F$ remains 1 until control exits from the $I$ loop and the test age $T$ becomes 20. In some cases there are two $X(I)$ values equal to the test age, resulting in the appropriate output 2 as shown, for example, for age 33. Finally, when all ages from 1 to 50 have been tested, control exits from the $J$ loop to END. There are 50 iterations of the $J$ loop, each of which causes 20 iterations of the $I$ loop; hence 20 x 50 = 1000 tests are made.

Because this program is the most difficult encountered so far, we sumarize its parts as follows. Dimension, data and read module are followed by nested loops, an outer loop that sequences through test ages 1–50 and an inner loop that tests each age against each datum in the data set. When correspondence is found, $F$ is incremented, hence $F$ acts as an accumulator of the frequency of $T = X(I)$. The frequency is printed out after which $F$ is reinitialized to zero and $T$ is incremented by 1 to test the next higher age.

### Exercise 3-5

Write a variation on Program 3-3 that prints out the frequency distribution of the integral data in Table 3-3 just as it appears in Table 3-4. (Hint: add another loop.)

## Cumulative Frequency Distributions

It is often of interest to know how many of a data set have a value that is greater than or less than some arbitrary boundary. In our example of parental ages, we might wish to know how many ages in the data set are greater than (or less than) 25, 30, 35, or some other age. Data arranged in this way constitute a *cumulative frequency distribution*, as in Table 3-5, which is derived from Table 3-4.

A data set arranged as this one is replies, at a glance, to questions like "how many of the parents surveyed were over 35?," with the answer 8. The graphical presentation of a cumulative frequency distribution is an *ogive*. An ogive is the figure resulting from a plot of cumulative frequency vs the quantity measured, in this case, age.

One of the uses of the ogive is in interpolation. Thus, from Fig. 3-1, if one inquires, "How many parents in the group surveyed were 27 or less?," the answer, as determined by the dotted line, is 5. If we look back at Table 3-3, we see that this answer is wrong and that there were, in fact, four parents under 27 in age. Thus, obtaining numerical data by interpola-

Table 3-5
Cumulative Frequency Distribution of the Data in Table 3-4

| Age | Number of parents: More than the designated age | Number of parents: Less than or equal to the designated age |
|---|---|---|
| 20 | 19 | 1 |
| 25 | 16 | 4 |
| 30 | 14 | 6 |
| 35 | 8 | 12 |
| 40 | 5 | 15 |
| 45 | 1 | 19 |
| 50 | 0 | 20 |

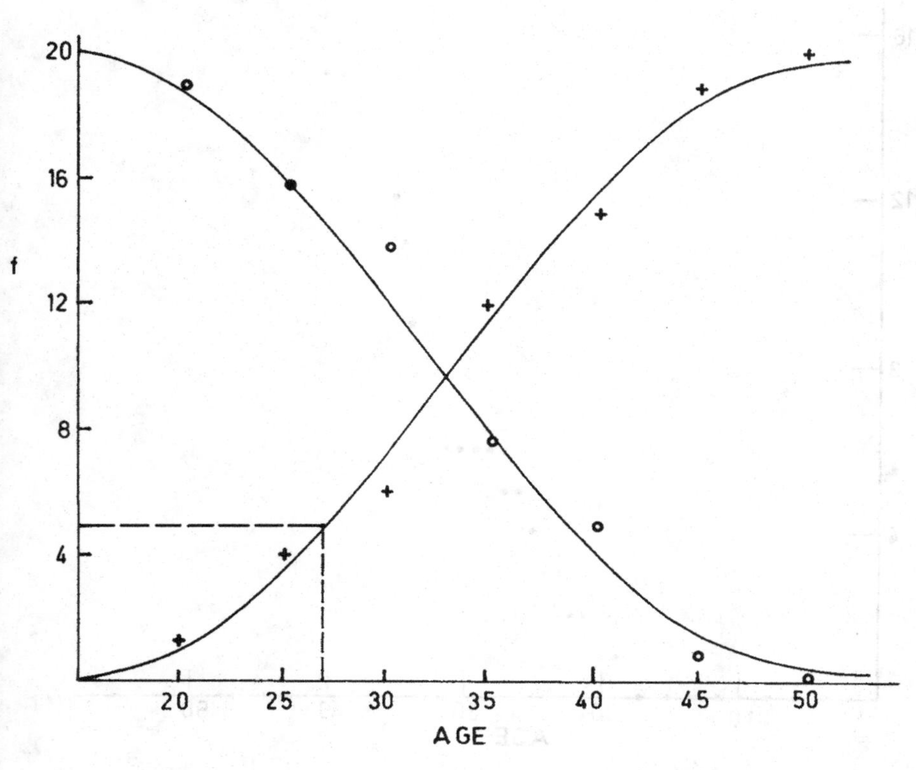


Fig. 3-1. Ogive resulting form the data in Table 3-5.

tion of a frequency ogive is an approximate method of estimating information that was present in the original data set, but has been lost in the process of data compression.

The value of a good and well-organized library of program listings is illustrated in the following exercise. You are asked to compute the *cumulative* frequency distribution, as contrasted to the computation in Program 3-3, which gives the frequency distribution. Sometimes very subtle changes in program logic can bring about large changes in the kind of output we obtain.

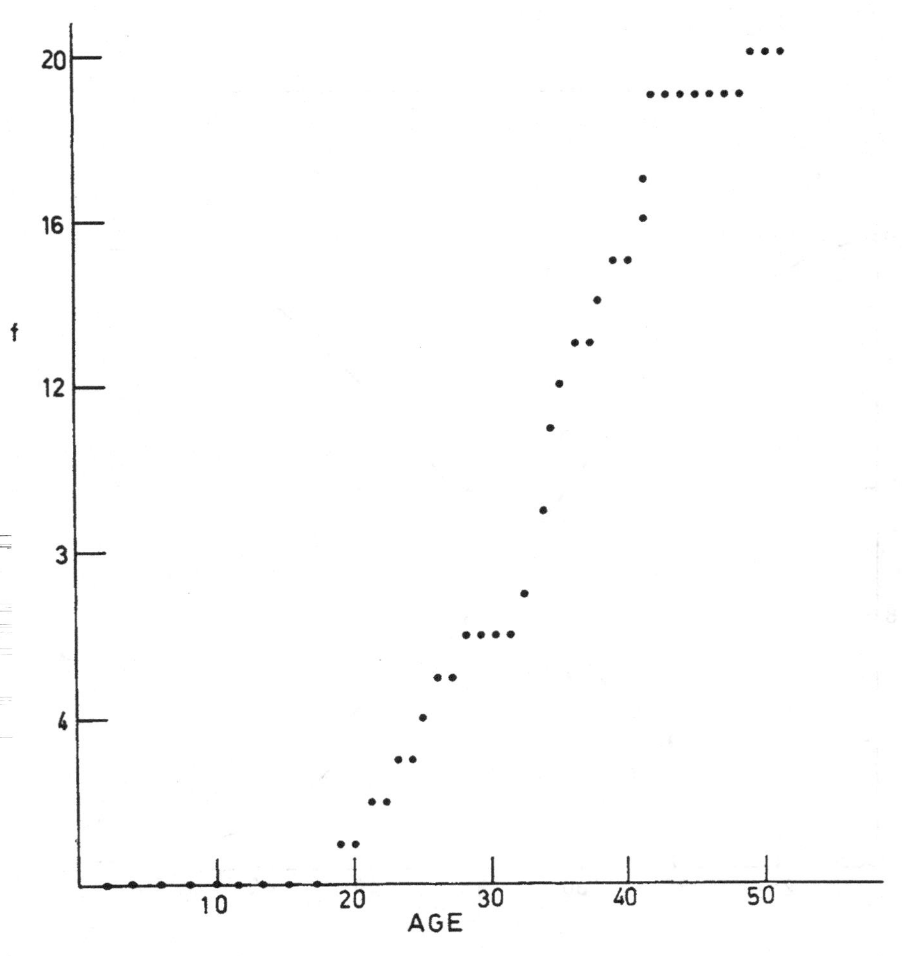


Fig. 3-2. Ogive resulting from the data in Table 3.3

### Exercise 3-6

Program 3-3 can be modified to compute the cumulative frequency distribution by merely exchanging program steps 60 and 70. Do so and run the program so as to obtain the cumulative frequency distribution for the data set contained in Table 3-3. Plot the results. The ogive obtained should be essentially the same as Fig. 3-2. Write a commentary on the new program explaining how the change from Program 3-3 has influenced the results.

## Application: Dose–Response Curve

It is important in pharmacology and medicine to know the effective dose, ED, in the case of a drug or the lethal dose, LD, of a toxic substance. Studies on the lethal dose of pesticides are important in agriculture, agricultural chemistry, and ecology. The statistical method used in ED studies is similar to that used in LD studies. We shall treat an experiment on the lethal dose of digitalis infused into a population of laboratory cats. Table 3-6 shows the minimum lethal dose of a solution of digitalis slowly administered to each of 220 etherized cats. At each increase of 0.05 mg of digitalis per kg body weight of the laboratory animal, the number of deaths is recorded.

Deaths within any 0.05 mg/kg dose interval are shown in column two of the table and the cumulative number of deaths at each total digitalis level is recorded in column three. When these cumulative data are plotted as a

Table 3-6
Minimum Lethal Dose of Digitalis
for 220 Laboratory Cats

| Dose, mg/kg | Deaths | Cumulative deaths |
|---|---|---|
| 0.40 | 2 | 2 |
| 0.45 | 4 | 6 |
| 0.50 | 8 | 14 |
| 0.55 | 22 | 36 |
| 0.60 | 26 | 62 |
| 0.65 | 52 | 114 |
| 0.70 | 50 | 164 |
| 0.75 | 24 | 188 |
| 0.80 | 20 | 208 |
| 0.85 | 6 | 214 |
| 0.90 | 4 | 218 |
| 0.95 | 2 | 220 |

$LD_{50}$ = 0.645 mg/kg

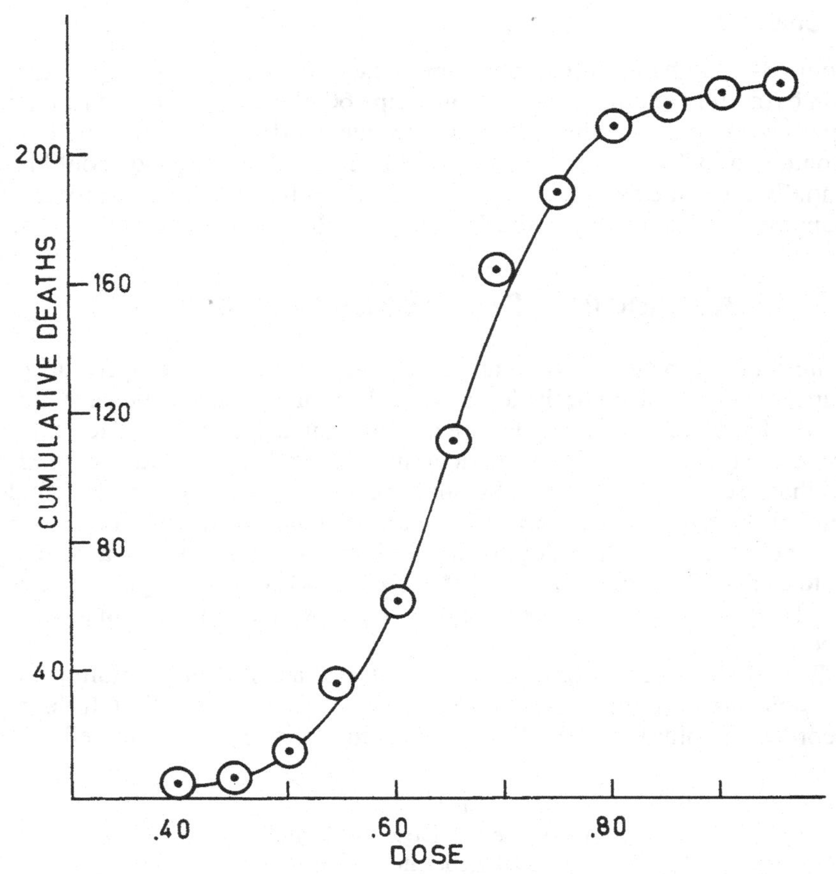

Fig. 3-3. Estimation of the $LD_{50}$ from a frequency ogive.

function of total dose, the ogive shown in Fig. 3-3 results. If the entire curve of a toxic substance being compared to digitalis lies to the right of the curve in Fig. 3-3, but has the same shape, the substance is not as toxic as digitalis. A useful parameter in comparing toxicities is the $LD_{50}$, i.e., the lethal dose for 50% of the population. This is obtained by reading up the vertical axis to 50% of the population, 110 in this case, constructing a horizontal so that it intersects the ogive and dropping a vertical to the dose axis. Figure 3-3 leads to an estimate of $LD_{50} = 0.65$ mg/kg animal weight.

Another property we learn from a dose–response curve is its steepness or variability. A steep dose–response curve indicates considerable sensitivity of response to dose, sometimes referred to as a narrow threshold.

## Frequency Grouping and Pictorial Representation of Continuous Data

The same methods are used in condensation and representation of continuous data as are used for integral data but some special problems are posed by the continuous variable. These will be illustrated by treating the data in Table 3-7.

We wish to write a frequency distribution for this data set, hence we must select an interval. The range of this data set is 45.7 − 36.1 = 9.6 cc, hence either a 1 or a 2 cc interval is a reasonable first choice. We shall select a 2 cc interval. The frequency distribution resulting from the set is shown in Table 3-8.

One difference we see between this distribution and the previous one is the splitting of a datum between two groups. The last datum in Table 3-7 happens to fall exactly on the boundary between two intervals, 38–40 and

Table 3-7
Cubic Centimeters of Cells per Hundred Cubic Centimeters of Blood for Normal Adults

| Individual | Cells, cc per 100 cc of blood | Individual | Cells, cc per 100 cc of blood |
|---|---|---|---|
| 1 | 39.2 | 11 | 40.7 |
| 2 | 43.0 | 12 | 39.2 |
| 3 | 42.5 | 13 | 41.4 |
| 4 | 37.7 | 14 | 41.3 |
| 5 | 39.2 | 15 | 38.2 |
| 6 | 45.0 | 16 | 40.3 |
| 7 | 44.9 | 17 | 41.9 |
| 8 | 36.1 | 18 | 39.9 |
| 9 | 45.7 | 19 | 42.1 |
| 10 | 41.2 | 20 | 40.0 |

Table 3-8
Frequency Distribution of the Data Contained in Table 3-7

| Cell volume, cc | Frequency |
|---|---|
| 36–38 | 2 |
| 38–40 | 5.5 |
| 40–42 | 6.5 |
| 42–44 | 3 |
| 44–46 | 3 |

40–42; hence a half frequency is given to one, and a half frequency to the other. This problem could have been handled in a different way by assigning the intervals differently. Had the intervals been set up 36–37.9, 38–39.9 etc., all data would have fallen unequivocally into one or another group.

There is some danger of skewing data by interval selection. For a large data set, if the intervals are selected in the way just described, each time a datum falls on what would normally be an interval boundary, it is placed in the higher group. In our example, the datum 40.0 would go into the group 40–41.9. If this is done many times, the entire distribution is displaced to the high side, as is the ogive. If intervals are selected, e.g., 36.1–38, 38.1–40, etc., apparent displacement takes place to the low side.

Cumulative frequency distributions and ogives are derived from frequency distributions of continuous data sets just as they were for integral data sets.

## Quartiles and Percentiles

Data sets are often described in terms of *quartiles* and *percentiles* and less frequently in terms of *deciles, quintiles,* or other arbitrary subdivisions. To illustrate, we shall consider the ideal ogive resulting from a large (strictly, infinite) data set with a perfectly random distribution.

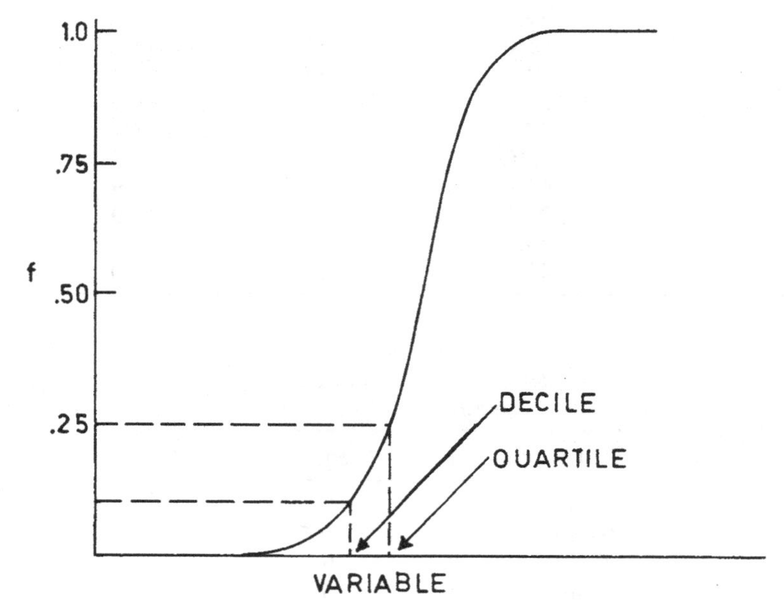


Fig. 3-4. Ideal ogive for a continuous or integral variable.

If we measure 1/10 of the way up the vertical axis, draw a horizontal to the ogive, and drop a line to the horizontal axis, we have that value of the variable below which lie one tenth of the data points and above which lie the remaining 9/10. This value of the variable is called the first decile. Measuring 2/10, 3/10, etc. of the way leads to the second, third, etc. deciles. If we measure 1/4 of the way up the vertical, we arrive at the first quartile; 1/2 of the way gives us the second quartile, and so on. Quintiles are measured in units of 1/5 and percentiles in units of 1/100. The quartile, decile, etc., are numbers so that it is not correct to say that a student scored in the third quartile of the class on an examination. A score might fall between the second and third quartile or, occasionally, it might be the third quartile. The second quartile is the point below which and above which one half of the data points lie, that is, it is the median. Ogives of small data sets do not approach the ideal curve of Fig. 3-4 very closely, hence the graphical method of selecting quartiles, and so on may not work. This is not a serious problem, however, because quartiles etc. may be obtained from a small set of data merely by inspection.

## Skewness and Kurtosis

When data points are normally distributed (perfectly random and infinite in number) the mean, median, and mode are identical. When the distribution is not perfectly random, the curve may be *skewed* to the right, as in Fig. 3-5, or to the left. A left-skewed curve would resemble the mirror image of Fig. 3-5. In these cases, the mean, median, and mode are not identical.

When a distribution is symmetrical about its vertical axis, but is more sharply peaked than a normal curve, it is said to be *leptokurtic*. When it is wider than a normal curve, it is said to be *platykurtic*. (The normal curve is called *mesokurtic*.) *Kurtosis* is said to exist if a distribution is either leptokurtic or platykurtic. Examples are shown in Fig. 3-6.

It is not necessary to compute the frequency distribution of a data set to determine its skewness or kurtosis. This information may be obtained from higher moments of the raw data set. The first moment about an arithmetic mean is

$$\pi_1 = \sum d_i/N \tag{3-13}$$

which is always zero as a previous exercise has shown. The second moment is

$$\pi_2 = \sum d_i^2/N \tag{3-14}$$

which is the variance and has been calculated for data sets above. The third moment is

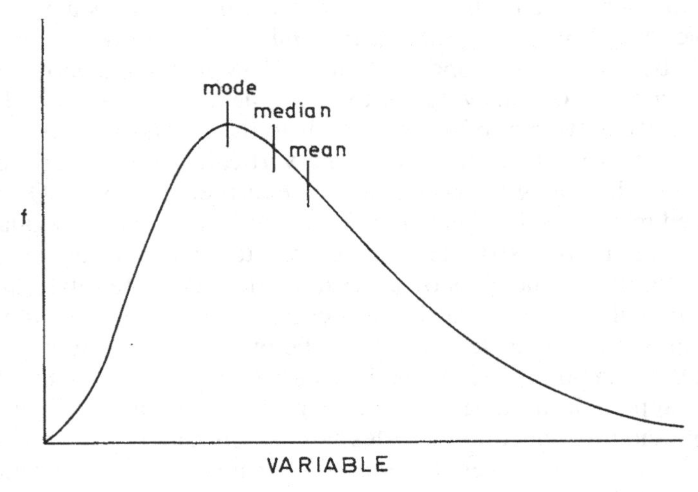


Fig. 3-5. Nonnormal frequency distribution skewed to the right.

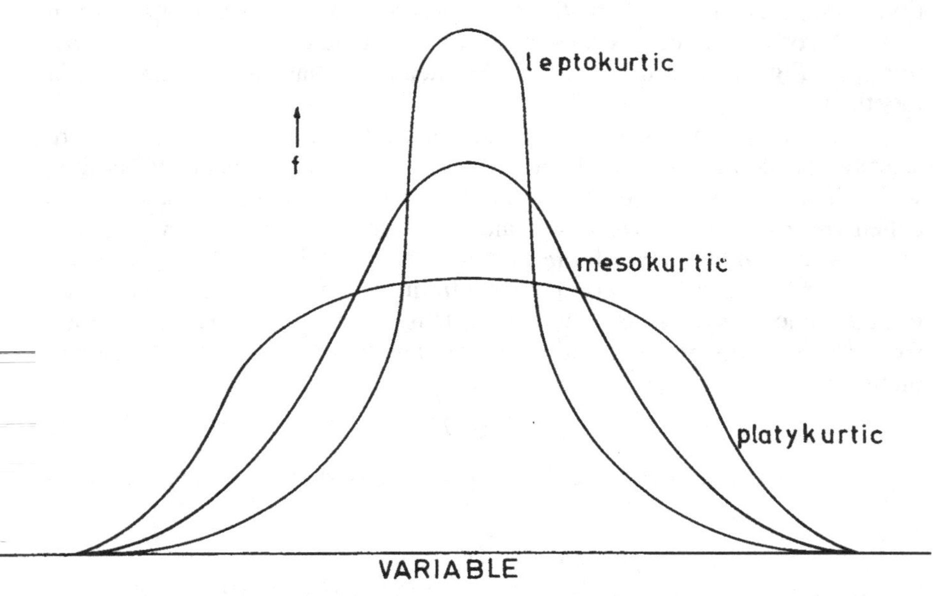


Fig. 3-6. Mesokurtic, leptokurtic, and platykurtic distributions.

$$\pi_3 = \sum d_i^3 \Big/ N \tag{3-15}$$

and the fourth is

$$\pi_4 = \sum d_i^4 \Big/ N \tag{3-16}$$

If the data set is symmetrical about its mean, $\Sigma d_i^3 = 0$, hence

$$\beta_1 = \pi_3^2 \Big/ \pi_2^3 \tag{3-17}$$

is also zero. When the data set is not symmetrical, $\beta_1 \neq 0$, which is a criterion of skewness, being large if the distribution is highly skewed and small if it is not. Further, $\pi_3$ indicates the direction of skewness, being positive for a data set skewed to the right and negative for one skewed to the left. Right- and left-handed skewness as used here assume that the data set is arrayed in increasing value from left to right.

A kurtosis parameter, $\beta_2$ is

$$\beta_2 = \pi_4 \Big/ \pi_2^2 \tag{3-18}$$

which is 3.0 for a mesokurtic curve. If $\beta_2 < 3.0$, the distribution is platykurtic and if $\beta_2 > 3.0$, it is leptokurtic.

## Exercise 3-7

Write a program to determine $\pi_1$, $\pi_2$, $\pi_3$, and $\pi_4$ for a data set containing $N$ variables. Use your program to treat the data set in Table 3-3 and make statements as to its skewness or lack of skewness and its kurtosis.

## Glossary

*Average Deviation.* Arithmetic mean of deviations, generally of measured values from their arithmetic mean.

*Coefficient of Variation.* Relative standard deviation expressed as a percentage.

*Cumulative Frequency Distribution.* Frequency distribution indicating how many members of a data set have a value that is greater than or less than some arbitrarily selected set of values over the range of the data.

*Decile.* Value of the independent variable corresponding to one, two,. . .tenths of the way up the cumulative frequency ogive.

*Dispersion.* Deviation of data from a central value; scatter.

*Frequency Distribution.* Data grouped in intervals according to the frequency of appearance of values within each interval.

*Frequency Group.* Subset of a data set falling within a prespecified interval.

*Inner Loop.* The loop of a nested pair that runs through one set of iterations for each iteration of an outer loop.

*Interval.* Subdivision of the range of a data set.

*Kurtosis.* Abnormality of the frequency distribution causing it to be peaked or flat.

*Leptokurtic Distribution.* An abnormally peaked frequency distribution.
*Mesokurtic Distribution.* A normal frequency distribution which is neither leptokurtic nor platykurtic.
*Nested Loops.* Program loops in which one loop runs through an entire set of iterations on each iteration of the other.
*Ogive.* S-shaped curve representing a cumulative frequency distribution of a random variable.
*Outer Loop.* The loop of a nested pair that, as part of each iteration, brings about an entire set of iterations of an inner loop.
*Percentile.* Value of the independent variable corresponding to one, two,. . .hundredths of the way up the frequency ogive.
*Platykurtic Distribution.* An abnormally flat frequency distribution.
*Population.* An entire set of people, objects, data, etc., possessing a common characteristic that differentiates it from other sets. Sometimes called a universe.
*Quartile.* Value of the independent variable corresponding to one, two, or three quarters of the way up the cumulative frequency ogive.
*Range.* The largest member of a data set minus the smallest.
*Sample.* A subset of a population smaller in number than the population.
*Skewness.* Degree to which a frequency distribution is distorted away from the symmetrical normal distribution.
*Standard Deviation.* Quadratic mean of deviations, generally of measured values from their arithmetic mean.
*Universe.* See population.

## Problems

*1.* What is the range of the data set: 5.4, 7.1, 7.2, 3.8, 5.1, 8.1, 6.5?

*2.* Prove (algebraically) that the arithmetic mean deviation is zero.

*3.* A group of 8 fresh water euglena were observed to have the following lengths in microns (1 micron = 0.001 mm): 37.1, 36.5, 30.0, 41.9, 40.1, 33.2, 37.7, 39.1. What are the arithmetic mean and standard deviation for this data set? Are these calculated values sample statistics or population parameters?

*4.* The data in problem 3 are used to estimate the mean length and standard deviation of euglena in the pond from which they were taken. Are these estimated values statistics or parameters of the population?

*5.* Determine the arithmetic mean of the data set 102, 105, 100, 99, 108, 107. Sum the deviations of this data set from their arithmetic mean and demonstrate that the sum is zero.

*6.* Determine the arithmetic mean and standard deviation of the data set in Table 3-7 using an appropriate computer program.

*7.* Determine the frequency distribution of the data in Table 3-7 using a 2 cc interval so as to obtain the data in Table 3-8.

*8*. Determine the frequency distribution of the data in Table 3-7 using a 1 cc interval.

*9*. Determine the greater than cumulative frequency distribution for the data in Table 3-7 using a 2 cc interval and plot the ogive.

*10*. Determine the less than cumulative frequency distribution for the data in Table 3-7 using a 1 cc interval and plot the ogive.

*11*. Modify Program 3-1 to provide a normal exit from the loop.

*12*. The output of Program 3-3 is too wide for some CRT screens. Investigate the use of a semicolon as punctuation in statement 120 to produce an output that is narrower but sacrifices no output characters.

*13*. If you are using a relatively slow machine, such as a home microcomputer, investigate the use of a delay loop, i.e., a loop that does nothing, to slow down the appearance of output data on the CRT Screen so as to make it easier to read. If you are timesharing, do not play with delay loops; they are expensive.

## Bibliography

A. K. Bahn, *Basic Medical Statistics*, Grune and Stratton, New York, 1972.

F. E. Croxton, *Elementary Statistics with Applications in Medicine and the Biological Sciences*, Dover, New York, 1953.

D. J. Koosis, *Statistics*, Wiley, New York, 1972.

# Chapter 4

# Understanding Probability

If one has the exact length and breadth of a rectangle, geometry permits one to calculate its area with certainty. If one wishes to calculate tomorrow's weather, or the diagnosis of a critically ill patient, no certain calculation is possible because the number of influencing factors is so large that one can know only a small fraction of them. With incomplete data, the answer is unknowable in any certain or absolute sense, but we are unwilling to throw up our hands and run away from such problems because we also know that whether we suffer famine or enjoy plenty depends largely on the weather, and that life or death often depends on a diagnosis. When necessary information for a desired calculation is unknown or unknowable, we can still calculate probable answers. The mathematics of probability has been devised to make these calculations as reliable as the incomplete input data permit them to be.

The probability, $p$, that an event will occur is the number of ways it can occur divided by the total number of ways all events can occur, provided that all events are equally likely. In flipping a coin, there is one way heads, H, can occur and one way tails, T, can come up. There is no event other than H or T that can occur. The total number of ways all events can occur is 2; hence the probability of flipping H is 1/2 and the probability of flipping T is also 1/2

$$p(\mathrm{H}) = p(\mathrm{T}) = 1/2 \tag{4-1}$$

A die has six faces; hence there are six events that can occur. One can throw a 1, 2, . . ., 6, but each event can occur in only one way. The probabilities of throwing 1, 2, . . ., 6 with one die are

$$p(1) = p(2) = p(3) = \ldots = p(6) = 1/6 \tag{4-2}$$

## Program 4-1

If the probability of rolling any point score, 1, 2, 3. . .6 with one die is 1/6, then the probable arithmetic mean of all the point scores is 1/6(1) + 1/6(2) . . . 1/6(6) = 21/6 = 3.500. Program 4-1 simulates one thousand random dice rolls and prints out the mean score.

```
10 FOR I=1 TO 1000
20 A=RND(6)
30 S=S+A
40 NEXT I
50 M=S/1000
60 PRINT M
70 END

READY
```

Commentary on Program 4-1. The new element in Program 4-1 is the random number generator RND that produces a random number between 1 and the number in parentheses, called the argument. In this simulation, the argument is 6 and RND(6) generates the numbers 1 through 6 just as rolling a die would do. The number or "score" generated on each simulated throw is accumulated in *S* over 1000 iterations or 1000 simulated throws. Statement 50 calculates the mean score which is printed by statement 60. We obtained 3.557, 3.515, 3.537, and 3.420 on four runs. By the nature of random numbers and probability, you should not expect to duplicate these results. You will be able to get closer to the theoretical value of 3.500, valid for an infinite number of rolls, if you increase the number of iterations (probably).

Some random number generators do not produce random integers as required by Program 4-1, but generate only decimal numbers between 0 and 1. Such a generator groups scores around 0.5 rather than 3.5. Problem 3 at the end of the chapter asks you to rewrite the program to accommodate this kind of random number generator.

## Simultaneous and Equally Acceptable Probabilities

When two coins are flipped simultaneously, the probabilities of H or T are independent, that is, the appearance of H or T for one has no influence on the other. These are *simultaneous independent events*. When the coins are flipped sequentially in time, the same laws apply; hence there is no such thing as a number that is "due" to come up in dice, roulette, or other games of pure chance. In this and all that follows, coins, dice, etc. are assumed to be "honest" unless stated or proven to be otherwise.

When two coins are flipped simultaneously or sequentially, there are four events that can occur, each in only one way, HH, HT, TH, and TT. The probability of flipping two heads is the number of ways HH can occur, namely one, divided by the number of ways all events can occur, $p(\text{HH}) = 1/4$. If we distinguish between the coins as "coin to the left, coin to the right" or, throwing them sequentially, "first coin, second coin," HT is a different event from TH. Each can occur in only one way hence, by the previous argument,

$$p(\mathrm{HH}) = p(\mathrm{HT}) = p(\mathrm{TH}) = p(\mathrm{TT}) = 1/4 \tag{4-3}$$

If we do not distinguish between the coins, an H and a T together constitute one event, but one that can occur two ways, HT or TH. Now,

$$p(\mathrm{HT\ or\ TH}) = 2/4 = 1/2 \tag{4-4}$$

We note from Eqs. (4-3) and (4-4) that $p(\mathrm{HH})$ is the product of $p(\mathrm{H})$ times itself. This is not coincidence; in general, the probability of $n$ events occurring simultaneously is the product of the individual probabilities

$$p(\mathrm{A\ and\ B\ and\ \ldots\ I\ and\ \ldots\ }N)$$

$$= p(\mathrm{A}) \times p(\mathrm{B}) \times \ldots p(N) = \prod^{N} p(\mathrm{I}) \tag{4-5}$$

In Eq. (4-5), the symbol $\prod^{N}$ *indicates that we must take the product of probabilities over all N* possible events.

Summing up all probabilities in Eqs. (4-3) or (4-4) either with or without distinguishing between coins, the sum is one: $1/4 + 1/4 + 1/4 + 1/4 = 1$ if we distinguish between HT and TH, and $1/4 + 1/2 + 1/4 = 1$ if we do not. This is not coincidence either, for the probability of any certainty is one. If we correctly sum up the probabilities of all possible events, it is certain that one of them will occur. Negative probabilities and probabilities greater than one do not exist, and probabilities equal to zero indicate that an event cannot occur, i.e., that its failure to occur is certain. If a calculated probability or the sum of calculated probabilities is less than zero or greater than one, an error has been made. If the sum of probabilities is greater than zero but less than one, it is likely that some possible event has been ignored and that the summation is incomplete.

If we throw three coins, there are eight possible events, each of which can occur one way. Going from a consideration of one to two, to three coins, we have the series 2, 4, 8 . . . for the total possible events. This is the geometric series $m^n$, where $m$ is the number of events that can occur for one trial, two in the case of coin flipping, and $n$ is the number of trials made, in our illustration, the number of coins flipped.

Consider the problem of rolling two dice and getting a 7. Rolling a single die can lead to six events; hence $m = 6$, and $n = 2$ for this problem. The total number of possible events is $6^2 = 36$. There are, however, six ways to roll a 7 with two dice. We do not distinguish between the dice; hence a 6 and a 1, or a 1 and 6, are both acceptable. So also are a 5 and 2, a 2 and 5, and other combinations represented by the solid lines in Table 4-1.

The probability of rolling a 7 is the number of ways that event can occur divided by the number of ways all events can occur

$$p(7) = 6/36 = 1/6 \tag{4-6}$$

We might wish to calculate the probability of a winning first roll in craps in which either a 7 or 11 wins. There are two ways of rolling an 11 as indicated by the dashed lines in Table 4-1. By the reasoning used above, $p(11) = 2/36 = 1/18$. The total number of ways either event can occur is $6 + 2 = 8$, hence the probability of either event occurring is $8/36 = 2/9$. But this is just equal to the sum $1/6 + 1/18 = 4/18 = 2/9$, which is the sum of $p(7) + p(11)$. Once again, this is not coincidence; in general, the probability of any one of a group of events occurring is the sum of the individual probabilities

$$p(\text{A or B or} \ldots \text{I or} \ldots N) = p(\text{A}) + p(\text{B}) + \ldots +$$

$$p(\text{I}) + \ldots p(N) = \sum^{N} p(\text{I}) \tag{4-7}$$

Now consider throwing two dice and a coin simultaneously (or sequentially without regard to order). Define a "win" as a 7 or 11 combination with the dice and a T with the coin.

$$p(7 \text{ or } 11) = 2/9$$
$$p(\text{T}) = 1/2 \tag{4-8}$$

We have seen that there are eight ways 7 or 11 can be thrown. There are, however, twice the number of possible events when we take into account the fall of the coin, for there are 36 possible dice combinations that may appear with H, and 36 more that may appear with T. By our definition of a win, *all* 36 possible dice rolls in combination with H are unfavorable, while 8 rolls in combination with T are favorable. The probability of a winning combination is

Table 4-1
Diagram Indicating Ways of Rolling Seven (Solid Lines) and Eleven (Dashed Lines) Using Two Dice

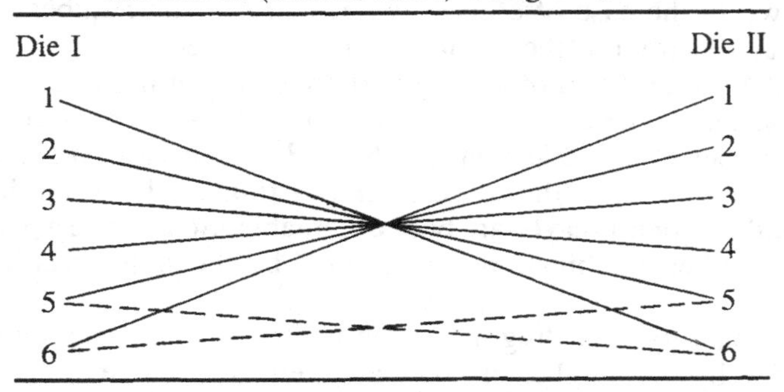

$$p(\text{win}) = 8/72 = 1/9 \qquad (4\text{-}9)$$

We arrive at the same conclusion using the rule that simultaneous probabilities for independent events must be multiplied to obtain the total probability

$$p(\text{win}) = p(7 \text{ or } 11) \times p(\text{T}) = 2/9 \times 1/2 = 1/9 \qquad (4\text{-}10)$$

or

$$p(\text{win}) = p(7 \text{ or } 11 \text{ and T}) = [p(7) + p(11)]p(\text{T}) = 1/9 \quad (4\text{-}11)$$

In general,

$$p[(\text{A or B}) \text{ and C}] = [p(\text{A}) + p(\text{B})]p(\text{C}) \qquad (4\text{-}12)$$

## Program 4-2

Let us test Eq. (4-11) by simulating the roll of two dice and the throw of one coin over many trials. If the equation is correct, the number of "wins" divided by the total number of tries should be 1/9 = 0.1111.

```
10 FOR I=1 TO 1000
20 A=RND(6)
25 A=6*Z
30 B=RND(6)
40 C=RND(2)
50 IF(A+B)=7 AND C=2 THEN S=S+1
60 IF(A+B)=11 AND C=2 THEN S=S+1
70 N=N+1
80 F=S/N
90 NEXT I
100 PRINT S,N,F
110 END

READY
```

**Commentary on Program 4-2.** By the rules of the game, 7 or 11 and T wins. Here we shall define C = 1 as designating a coin toss of heads and C = 2 as a coin toss of tails. Thus A simulates one die, B simulates the other and C simulates the coin. The new feature of this program is an expanded form of the IF statements, 50 and 60. They are nearly self-explanatory as they stand. For example, 50 translates in the terminology of the game to "If the sum of spots on the dice is 7 and the coin is T, then add one to the number of wins, S." Statement 60 increments the win tally for the combination 11, T. Statement 70 is a counter that accumulates the number of times the game is played and F is the frequency of wins relative to total games. This program takes 10 to 15 min to run on a microcomputer (1.78 MHz). Actual runs gave F = 0.1111, 0.1103, 0.1017 and 0.1145. Fluctuations like these are the reason people gamble. Some systems will not run statement 50 as written. Problem 4 asks you to break statements 50 and 60 into separate IF. . .THEN statements.

## Exercise 4-1

What is the probability of cutting to a nine or a spade in a normal, well-shuffled deck of 52 cards?

Solution 4-1. The probability of cutting to a nine is 1/13, and that of cutting to a spade is 1/4, which might lead us to suppose that the probability is

$$p(9 \text{ or spade}) = p(9) + p(\text{spade}) = 1/13 + 1/4 = 0.327(?!)$$

If we do this, we commit the error of counting the nine of spades twice, once as a nine and once as a spade. The true probability is the number of ways of drawing a nine *or* a spade divided by the total number of probable draws

$$p(9 \text{ or spade}) = \frac{(4 + 13) - 1}{52} = 0.308$$

or

$$p(9 \text{ or spade}) = 1/13 + 1/4 - 1/52 = 0.308$$

## Application: Heredity

Phenylketonuria (PKU) is an inherited metabolic defect that is characterized by mental retardation. The PKU gene is recessive and the PKU syndrome is determined by one gene such that each individual falls into one of four genotypes,

$$XX \qquad Xx \qquad xX \qquad xx$$

where $X$ is the dominant gene and $x$ is the recessive. $XX$ and $xx$ genotypes are called homozygous; $Xx$ and $xX$ are heterozygous. Genotypes $Xx$ and $xX$ are not distinguishable from each other and are not distinguishable from genotype $XX$. None is PKU afflicted because the dominant gene, $X$, suppresses the recessive PKU gene. Only when both genes inherited from the parents are PKU genes does the the syndrome appear. Offspring of $XX$ parents are normal, offspring of an $Xx$ and an $XX$ parent are normal, those of two $Xx$ parents or offspring of one $Xx$ and one $xx$ parent may be normal or afflicted.

In many inherited diseases of this kind, the affliction is so severe that $xx$ individuals do not reproduce. If two $Xx$ parents reproduce, the offspring receives one gene from each parent, randomly selected from the four genes available. The probabilities are the same as those of coin-tossing,

$$p(XX) = p(Xx) = p(xX) = p(xx) = 1/4 \qquad (4\text{-}13)$$

hence the probability of a PKU child is 1/4. The parents, although normal, are called *carriers* because they carry one recessive PKU gene that may or may not be transmitted to the children.

It is not possible to distinguish a recessive gene carrier from an $XX$ genotype because of the dominant gene he or she has. It is, however, possible to determine the probability that an individual may be a carrier on the grounds of his or her genetic history. Suppose a normal individual has a PKU sibling. We wish to determine the probability that the individual will have PKU offspring.

Since the sibling was of the $xx$ genotype, both parents, presumably normal, must have been carriers. The probability that the individual is a carrier is 2/3 because one genotype, $xx$, has already been ruled out by his normalcy. This leaves only three possibilities, $XX$, $Xx$, and $xX$. The PKU syndrome being rare, it is reasonable to assume that the individual's mate will be a normal homozygote and that the probability of PKU offspring is vanishingly small.

Suppose, however, that both parents have PKU siblings, what happens to the probability of PKU offspring? By the reasoning above, each parent has a 2/3 probability of being a carrier. By the rule of simultaneous probabilities, the probability that both male and female parents are carriers is the product

$$p(\text{M and F}) = p(\text{M})p(\text{F}) = 2/3 \times 2/3 = 4/9$$

The probability that the offspring will suffer the PKU trait is 1/4 if both parents are carriers. The sequential probability that (a) both parents are carriers and (b) the child is homozygous $xx$ is the product

$$p(\text{M and F and } xx) = 4/9 \times 1/4 = 4/36 = 1/9$$

which is the probability that the offspring will be PKU afflicted.

### Exercise 4-1

What is the probability that the offspring of parents both having PKU siblings will be PKU carriers?

## The Urn Problem: Dependent Probabilities

Problems involving sampling are often illustrated by the classical urn problem. Consider an urn containing a number of objects. The model represents a *population* about which we would like to know some characteristics. If the population in the urn consists of red and green marbles, we might wish to know how many are red and how many are green. If the population is a healthy group of people who have been subjected to infection by diphtheria, we might wish to know how many are likely to have contracted the disease and how many are not.

Following our simple model of the urn, suppose it contains a population of 20 objects, seven of which are red and 13 of which are green. We can define success as the act of drawing an object of either color; suppose we define drawing a red object as success. On one draw, there are twenty events that might occur. Seven of these are successes and thirteen are not. The probability of scoring a success is

$$p(\mathrm{S}) = 7/20 = 0.35 \tag{4-16}$$

Now, suppose we would like to redefine success as drawing a red object and a green object without distinction as to order. There are two ways of achieving success, R and G or G and R. There are also two ways of drawing the objects, one without replacement and one with replacement.

Suppose we draw the first object and do not replace it. We then draw the second object. As noted, our probability of success in drawing a red object first is 0.35. Our probability of drawing the second object and finding it green must be augmented if we have found a red object on the first trial since our success depleted the red population and increased the green population relative to it. The probability of drawing a green on the second try, having drawn and not replaced a red on the first try, is 13/19. The 19 in the denominator is the number of objects remaining in the urn after one has been withdrawn. The probability of drawing a red on the first trial followed by a green on the second is the product of the individual probabilities,

$$p(\mathrm{RG}) = 7/20 \times 13/19 = 91/380 = 0.239 \tag{4-17}$$

If we are not concerned with the order of drawing, any combination of a red and a green constitutes a success. Hence $p(\mathrm{GR})$ is just as interesting as $p(\mathrm{RG})$ and, by the rule of equally acceptable probabilities, we must sum $p(\mathrm{GR})$ and $p(\mathrm{RG})$ to obtain the probability of a success on both trys. The probability of scoring a green followed by a red is

$$p(\mathrm{GR}) = 13/20 \times 7/19 = 91/380 = 0.239 \tag{4-18}$$

and the probability of a success is the sum

$$p(\mathrm{S}) = p(\mathrm{RG}) + p(\mathrm{GR}) = 91/380 + 91/380 = 0.479 \tag{4-19}$$

Suppose we draw from our urn containing seven red objects and thirteen green ones, replace the object, mix, and then draw again. The probability of drawing a red object is as it was before, 0.35. With replacement, however, the probability of drawing a green object on the second draw is not augmented because the number of possible draws has not been decreased. Qualitatively, we expect the total probability of success to be less favorable than it was without replacement. Now,

$$p(\mathrm{RG}) = 7/20 \times 13/20 = 91/400 = 0.2275 \qquad (4\text{-}20)$$

Similarly,

$$p(\mathrm{GR}) = 13/20 \times 7/20 = 91/400 = 0.2275 \qquad (4\text{-}21)$$

The total probability of success is the sum of the individual probabilities, which is 0.455, slightly less than the total probability of success without replacement, as expected.

The reason these two probabilities differ is that drawing two objects from a finite population without replacement does not constitute a situation of simultaneous independent probabilities. Withdrawal of an object of either color changes the relative populations of colored objects and influences the outcome of the second draw. As the population becomes larger, the magnitude of this dependence becomes smaller until, in the limiting case of the infinite population, withdrawal of one object does not change the relative populations and dependence of the second draw on the outcome of the first disappears. Withdrawal with replacement does not change the relative populations either; hence it follows the same rules as the infinite population.

## Conditional Probabilities

In the section on sampling without replacement, we used the idea that the probability of a sequence of two dependent events is the probability of the first event times the probability of the second event, assuming a knowledge of the first outcome. Probabilities of events assuming that some other event has already occurred are called *conditional probabilities*. We shall introduce the notation $p(b \mid a)$ as the probability of $b$ after $a$ is known to have taken place. Using this notation, Eq. (4-17) may be written

$$p(\mathrm{RG}) = p(\mathrm{R}) \times p(\mathrm{G} \mid \mathrm{R}) \qquad (4\text{-}22)$$

and Eq. (4-18) becomes

$$p(\mathrm{GR}) = p(\mathrm{G}) \times p(\mathrm{R} \mid \mathrm{G}) \qquad (4\text{-}23)$$

where $p(\mathrm{RG})$ is the probability of drawing a red followed by a green and $p(\mathrm{GR})$ is the probability of drawing a green followed by a red. It was also seen that $p(\mathrm{RG}) = p(\mathrm{GR})$ hence

$$p(\mathrm{R}) \times p(\mathrm{G} \mid \mathrm{R}) = p(\mathrm{G}) \times p(\mathrm{R} \mid \mathrm{G}) \qquad (4\text{-}24)$$

Rearranging Eq. (4-24), we have

$$p(\mathrm{G} \mid \mathrm{R}) = \frac{p(\mathrm{G}) \times p(\mathrm{R} \mid \mathrm{G})}{p(\mathrm{R})} \tag{4-25}$$

with a similar expression for $p(\mathrm{R} \mid \mathrm{G})$.

Both of the above are restricted forms of *Bayes' theorem*.

## Example 4-2

Substitute the appropriate values on the right of Eq. (4-24) to demonstrate that this restricted form of Bayes' theorem yields $p(\mathrm{G} \mid \mathrm{R})$, i.e., the probability of drawing G having already drawn R without replacement which we already know to be 13/19 = 0.684. Repeat for $p(\mathrm{R} \mid \mathrm{G})$.

Solution 4-2. From Eq. (4-22)

$$p(\mathrm{G} \mid \mathrm{R}) = \frac{(13/20) \times (7/19)}{7/20} = 0.684$$

Similar treatment leads to

$$p(\mathrm{G} \mid \mathrm{R}) = \frac{(7/20) \times (13/19)}{13/20} = 0.368$$

Both these values appear as intermediate results in Eq. (4-17) and (4-18).

## Bayes' Theorem

Switching to a more general notation for Bayes' theorem as we have developed it up to this point,

$$p(b \mid a) = \frac{p(b) \times p(a \mid b)}{p(a)} \tag{4-26}$$

where $p(a)$ and $p(b)$ are the independent probabilities of events a and b and $p(a \mid b)$ and $p(b \mid a)$ are the conditional probabilities of $a$ given $b$ and of $b$ given $a$.

Suppose, however, that there are a number of events, $b_i$, that are not $a$. Now, $a$ can occur in combination with any one of them and

$$p(a) = \sum p(b_i) \times p(a \mid b_i) \tag{4-27}$$

If we want to calculate the probability of the combination of $a$ with one specific event, $b_i$, substitution of Eq. (4-27) into Eq. (4-26) gives

$$p(b_i \mid a) = \frac{p(b_i) \times p(a \mid b_i)}{\sum p(b_i) \times p(a \mid b_i)} \qquad (4\text{-}28)$$

which is a more general form of Bayes' theorem. The denominator contains the sum of probabilities of all possible combinations of events $b_i$ with $a$, and the numerator contains the probabilities of one specific combination, $b_i$ and $a$.

To better understand this powerful theorem, consider the problem of three urns, one containing two red balls, one containing a red and a green ball, and the third containing two green balls. Suppose, not knowing which urn is which, we draw a red from one of the urns without replacement. What is the probability of drawing a red from the same urn *on the second draw?* Clearly, the urn we have chosen to draw from is not the urn with two greens. The urn we have drawn from must be the urn with a red and a green or the one with two reds. Hence the probability of drawing a red ball on the second draw is the number of remaining red balls divided by the total number of balls, or 2/3.

When we apply Bayes' theorem to this problem, we begin to appreciate some of its subtleties and to anticipate some of its applications. Let us rephrase the problem as follows: having drawn a red ball, *the probability that we shall draw a red ball on the second draw is the same as the probability that we have initially selected the urn containing two red balls.*

Let us make another slight change in phraseology by making the hypothesis that the urn we are drawing from is the RR urn. Let us denote this hypothesis by $H_{RR}$. Initially our probability of selecting the RR urn from among the three was 1/3, but having drawn one red ball, we *know something* about the urn that we didn't know before; it is not the GG urn. Our probability of success is augmented; hence $p > 1/3$. Now the probability of drawing a second red ball is the same as the probability that the hypothesis just stated is true. Let us denote the probability that the hypothesis $H_{RR}$ is true as $p(H_{RR})$. Now, the question asked in the three-urn problem has been cast in a form which asks us to find the conditional probability $p(H_{RR} \mid R)$, i.e., the probability that the hypothesis "we are drawing from the RR urn" is true given that the result of the first draw was R. By Bayes' theorem

$$p(H_{RR}|R) = \frac{p(H_{RR})p(R|H_{RR})}{p(H_{RR})p(R|H_{RR}) + p(H_{RG})p(R|H_{RG}) + p(H_{GG})p(R|H_{GG})} \qquad (4\text{-}29)$$

We can evaluate all terms on the right from the stipulations of the problem. Since there are three urns, the hypothesis that our initial selection was the RR urn is as likely as either of the remaining two hypotheses,

$$p(\mathrm{H_{RR}}) = p(\mathrm{H_{RG}}) = p(\mathrm{H_{GG}}) = 1/3 \tag{4-30}$$

The probability that the initial draw will be a red ball *given* the hypothesis that we are drawing from the RR urn is one:

$$p(\mathrm{R} \mid \mathrm{H_{RR}}) = 1 \tag{4-31}$$

given the hypothesis that we are drawing from the RG urn, the conditional probability of a red is

$$p(\mathrm{R} \mid \mathrm{H_{RG}}) = 1/2 \tag{4-32}$$

and given that we are drawing from the GG urn,

$$p(\mathrm{R} \mid \mathrm{H_{GG}}) = 0 \tag{4-33}$$

Substituting into Bayes' theorem,

$$p(\mathrm{H_{RR}} \mid \mathrm{R}) = \frac{(1/3)(1)}{(1/3)(1) + (1/3)(1/2) + (1/3)(0)} = \frac{1/3}{3/6} = 2/3 \tag{4-34}$$

which, as we have seen is the probability of drawing a second red ball from the urn.

The importance of this problem is not in calculating the probability $p(\mathrm{R} \mid \mathrm{R})$, which we have already determined by inspection, but in revealing Bayes' theorem as a means of calculating the effect of an observation (experiment, test) on a hypothesis concerning the system.

If we make a certain hypothesis and perform a test the outcome of which agrees with the hypothesis, we are more confident in the hypothesis than we were before the test. We judge the probability that the hypothesis is true to be higher after the test than we did before it. The opposite is true if the test results are negative. Bayes' theorem enables us to see how much the probability of a correct hypothesis has been changed by the test. An example is given in the following application to diagnosis.

## Application of Bayes' Theorem: Diagnosis

Suppose, by the history and symptoms, that a patient is thought to suffer from disease $x$, $y$, or $z$ with the probabilities

$$p(x) = 0.50 \qquad p(y) = 0.20 \qquad p(z) = 0.30 \tag{4-35}$$

The patient is subjected to a test that gives a positive result 10% of the time when administered to a patient with disease $x$, 80% of the time for disease $y$, and 30% of the time for disease $z$. The test gives a positive result. How are the probabilities of diseases $x$, $y$, and $z$ changed; i.e., how is the diagnosis changed?

A diagnosis is a hypothesis that "the patient suffers from disease $x$ (or $y$ or $z$)." Let $p(t)$ represent the probability that the test is positive. The previous paragraph translates into the conditional probabilities

$$p(t \mid x) = 0.10 \qquad p(t \mid y) = 0.80 \qquad p(t \mid z) = 0.30 \quad (4\text{-}36)$$

We wish to calculate the new probabilities of diseases $x$, $y$, and $z$, given the positive test, $t$, i.e., we want the conditional probabilities $p(x \mid t)$, $p(y \mid t)$, and $p(z \mid t)$. Unlike the urn problem, these new conditional probabilities are not obvious and cannot be obtained by inspection. We use Bayes' theorem

$$\begin{aligned} p(x \mid t) &= \frac{p(x)p(t \mid x)}{p(x)p(t \mid x) + p(y)p(t \mid y) + p(z)p(t \mid z)} \\ &= \frac{0.50(0.10)}{0.50(0.10) + 0.20(0.80) + 0.30(0.30)} \\ &= \frac{0.050}{0.30} = 0.17 \end{aligned} \qquad (4\text{-}37)$$

Similar calculations for $y$ and $z$ yield

$$p(y \mid t) = 0.53 \qquad p(z \mid t) = 0.30 \qquad (4\text{-}38)$$

Although the conditional probability of disease $z$, given positive test result $t$, has not changed, there has been a large change in the conditional probabilities of diseases $x$ and $y$.

Suppose we apply the same test again or one with the same conditional probabilities, $p(t \mid x)$, $p(t \mid y)$, and $p(t \mid z)$. Suppose further, that the test is positive again. Now we can recalculate the conditional probabilities of diseases $x$, $y$, and $z$ to obtain $p'(x \mid t)$, $p'(y \mid t)$, and $p'(z \mid t)$. To do this, we replace our first set of probabilities, 0.50, 0.20, and 0.30, with the conditional probabilities obtained from the first test, 0.17, 0.53, and 0.30. Now,

$$p'(x \mid t) = \frac{0.17(0.10)}{0.17(0.10) + 0.53(0.80) + 0.30(0.30)} = 0.032 \qquad (4\text{-}39)$$

Similarly,

$$p'(y \mid t) = 0.80 \qquad p'(z \mid t) = 0.17 \qquad (4\text{-}40)$$

Our diagnosis has changed again owing to testing, and Bayes' theorem has enabled us to make quantitative calculations of the changes that have taken place. Bayes' theorem shows quantitatively how conditional probabilities change as we gain more information about a system.

## Permutations

The permutations of a data set are the number of ways in which the data may be arranged. Thus, the numbers 1 and 2 may be arranged 12 or 21 and the numbers 1, 2, and 3 may be arranged 123, 231, 312, 213, 321, and 132.

The previous examples have started out the sequence 2, 6, 24, . . . which is given by the factorial of the number of things arranged. The factorial of a number is denoted by an exclamation point.

$$\begin{aligned} 1! &= 1 \\ 2! &= 2 \times 1 = 2 \\ 3! &= 3 \times 2 \times 1 = 6 \\ 4! &= 4 \times 3 \times 2 \times 1 = 24 \end{aligned} \qquad (4\text{-}14)$$

### Exercise 4-3

Suppose we wish to arrange a group of 10 experimental mice into distinguishable groupings. How many arrangements are possible?

Solution 4-3. Arrangements of any ten distinguishable objects are given by the number of permutations of the number 10.

$$10! = 10 \times 9 \times 8 \times \ldots \times 1 = 3{,}628{,}800$$

### Program 4-3

```
1    REM PROGRAM TO FIND N! WHERE N GOES FROM 1 TO 20
10   LET N=0
20   LET F=1
30   LET N=N+1
40   LET F=F*N
50   PRINTUSING 60,N,F
60   :        ##! = ##.####^^^^
70   IF N>=20 GOTO 90
80   GO TO 30
90   END

READY
RUNNH


       1! =   1.0000E+00
       2! =   2.0000E+00
       3! =   6.0000E+00
       4! =   2.4000E+01
       5! =   1.2000E+02
       6! =   7.2000E+02
       7! =   5.0400E+03
       8! =   4.0320E+04
       9! =   3.6288E+05
      10! =   3.6288E+06
```

```
11! =  3.9917E+07
12! =  4.7900E+08
13! =  6.2270E+09
14! =  8.7178E+10
15! =  1.3077E+12
16! =  2.0923E+13
17! =  3.5569E+14
18! =  6.4024E+15
19! =  1.2165E+17
20! =  2.4329E+18

TIME:   0.60 SECS.
```

### Exercise 4-4

Write the commentary for Program 4-3. The PRINTUSING statement designates an *output format,* in this case, a two digit number followed by ! =, followed by a six digit exponential number.

## Combinations

The method used in Exercise 4-3 is general. The number of ways in which any number of objects can be arranged is the factorial of the number of objects taken.

Not infrequently, we wish to arrange a number of objects in groups of a certain size. Suppose that we choose to arrange 4 experimental mice in groups of 2. We are not concerned with the arrangement of mice within any group, but we wish to know how many different groupings are possible.

From among mice *a*, *b*, *c*, and *d*, we have 4 choices as to which mouse is selected first and three choices for the second selection. When we have selected two mice, we have no other options. The total number of selections is shown in Table 4-2.

The first choice can be *a*, *b*, *c*, or *d*, while, if the first choice was *a*, the second choice is limited to three possibilities, *b*, *c*, and *d* and so on for other possible choices. From Table 4-2, we see that not all choices are really distinguishable. Selecting *a* followed by *b* gives the same pairing as

Table 4-2
Total Number of Ways of Selecting Two Experimental Mice From a Group of Four

| | | | |
|---|---|---|---|
| *ab* | *ba* | *ca* | *da* |
| *ac* | *bc* | *cb* | *db* |
| *ad* | *bd* | *cd* | *dc* |

selecting $b$ followed by $a$. There are six sets of identical pairings in Table 4-2, *ab–ba*, *ac–ca*, etc.; hence the nunber of distinguishable pairings is reduced to six, *ab*, *ac*, *ad*, *bc*, *bd*, and *cd*. Another way of looking at the problem is to note that there are $4 \times 3 = 12$ permutations, each having two identical arrangements. This leads to $12/2 = 6$ distinct pairings.

If we wish to arrange 10 experimental mice in groups of 4, there are 10 ways we can choose the first mouse, 9 ways we can choose the second, 8 possible choices for the third, and 7 for the fourth. Since each group can contain only four mice, we have run out of options. We have $10 \times 9 \times 8 \times 7$ groupings of four mice in this set just as we had $4 \times 3$ pairings of four mice in the previous one. As we saw before, however, not all groupings are distinct and we must divide the total number of groupings by the number of indistinguishable ways of arranging a group of four mice. That number of ways is just the number of permutations of four objects, 4!. The number of *distinct* groupings possible is called the number of *combinations* for the set and is denoted

$$\binom{N}{n}$$

For 10 mice taken in groupings of 4, the number of combinations is

$$\binom{N}{n} = \binom{10}{4} = \frac{10 \times 9 \times 8 \times 7}{4 \times 3 \times 2 \times 1} = \frac{5040}{24} = 210 \qquad (4\text{-}42)$$

In general, when we want to group $N$ objects in groups of $n$ we have $N$ ways of making the first choice, $N - 1$ ways of making the second choice, and so on down to our last choice, which fills the group with $n$ objects and can be made in $N - n + 1$ ways. In our example above, $N = 10, N - 1 = 9, \ldots, N - n + 1 = N - 3 = 7$. The general formula for the number of combinations of $N$ things taken $n$ at a time is

$$\binom{N}{n} = \frac{N(N-1)(N-2)\ldots(N-n+1)}{n!} \qquad (4\text{-}43)$$

It is convenient to multiply both numerator and denominator in Eq. (4-43) by $(N - n)!$. This completes the series in the numerator and makes it equal to $N!$.

$$N(N-1)(N-2)\ldots(N-n+2)(N-n+1)(N-n)$$
$$(N-n-1)\ldots(2)(1) = N! \qquad (4\text{-}44)$$

Now,

$$\binom{N}{n} = \frac{N!}{(N-n)!n!} \qquad (4\text{-}45)$$

which is a more conventional form than Eq. (4-43).

## Program 4-4

The expression for $N$ things taken $n$ at a time in Eq. (4-45) is very important in the probability theory, to be developed in the next chapter, of things that can happen in either one of two ways, heads or tails, male or female, survive or die, etc. It is called the *binomial coefficient*. Program 4-4 calculates the number of combinations of $N$ things $M$ at a time. Note the slight change in notation in the BASIC program caused by the lack of lower case letters in many versions of BASIC.

```
1   REM PROGRAM TO FIND THE NUMBER OF COMBINATIONS OF N THINGS TAKEN M
2   REM AT A TIME
10  DIM A(3),B(3)
20  INPUT N, M
30  A(1)=N
31  A(2)=M
32  A(3)=N-M
40  FOR I=1 TO 3
50  J=0
51  F=1
60  J=J+1
70  F=F*J
80  IF J=A(I) GO TO 100
90  GO TO 60
100 B(I)=F
110 NEXT I
120 C=B(1)/(B(2)*B(3))
130 PRINT"THERE ARE"C"COMBINATIONS OF"N"THINGS TAKEN"M"AT A TIME"
140 GO TO 20
150 END

READY
RUNNH


?8,1
THERE ARE 8 COMBINATIONS OF 8 THINGS TAKEN 1 AT A TIME
?8,2
THERE ARE 28 COMBINATIONS OF 8 THINGS TAKEN 2 AT A TIME
?8,3
THERE ARE 56 COMBINATIONS OF 8 THINGS TAKEN 3 AT A TIME
?8,4
THERE ARE 70 COMBINATIONS OF 8 THINGS TAKEN 4 AT A TIME
?8,5
THERE ARE 56 COMBINATIONS OF 8 THINGS TAKEN 5 AT A TIME
?8,6
THERE ARE 28 COMBINATIONS OF 8 THINGS TAKEN 6 AT A TIME
?8,7
THERE ARE 8 COMBINATIONS OF 8 THINGS TAKEN 7 AT A TIME
?
```

**Commentary on Program 4-4.** Statements 1 through 20 are remarks, and dimension and input statements, that should be familiar. Statements 30, 31, and 32 store $N$, $M$, and $N - M$ in locations A(1), A(2), and A(3). Statement 40 enters a three-iteration FOR–NEXT loop that is terminated by statement 110. Statements 50 and 51 initialize J and F at zero and one, and statement 60 increments J to one on the first iteration while statement

70 stores, in location F, the product of the value already at that location times J. On the first iteration, F = 1 is merely replaced by F*J which is also 1. Statement 80 tests to find out whether J is equal to A(1), which is $N$, the number of things taken and if it is, skips statement 90. If J is less than A(1), which it will be on the first iteration of a program under the normal circumstance that $N$ is greater than one, statement 90 is executed in normal sequence and returns us to statement 60. Statements 60 and 90 constitute an *inner* loop *nested* in the *outer* FOR–NEXT loop. On the second iteration of the inner loop, J is incremented by one, multiplied into the value stored at F, with the product of this multiplication then being stored as the new value at F. Location F now has $1 \times 2$ as its stored value.

After testing to see whether J has been incremented to equal $N$ or not, the loop may be iterated again to store $1 \times 2 \times 3$ at F, and so on. When J finally equals $N$, $N!$ is stored at location F. Statement 80 causes an exit from the inner loop to 100, which stores $N!$ in location B(1).

Statement 110 causes a return to the beginning of the outer loop at 40 whereupon the entire process described above is repeated to find the factorial of the number $M$ at location A(2). The process is repeated on the third execution of the outer loop with the end result that $N!$, $M!$, and $(N - M)!$ are stored at locations B(1), B(2), and B(3), respectively. After exit from the loop,

$$\mathrm{B(1)/B(2)B(3)} = N!/M!(N - M)! \tag{4-46}$$

is stored at location C and printed out by an appropriate PRINT module.

Statement 140 is an unconditional GO TO statement and rereadies the system for new input values of $N$ and $M$. As written, the program constitutes an infinite loop and the END statement is never executed. A program of this type is suited to teletype terminal use because the operator must decide after each number of combinations of $N$ things taken $M$ at a time whether to input new values of $N$ and $M$ or to intentionally interrupt the loop and abort the program. Do not batch process an infinite loop.

## The Binomial Coefficient

Consider the expansion of the binomial expression, $(x + y)^3$

$$(x + y)^3 = x^3 + 3x^2y + 3xy^2 + y^3 \tag{4-47}$$

The coefficients of this expression, 1, 3, 3, 1, can be determined using the binomial coefficient equation, Eq. (4-45). When we look at the group $x^3 = xxx$, we are taking 3 things 3 at a time. There is only one way we can take this group which is shown by the value of its binomial coefficient

$$\binom{N}{n} = \binom{3}{3} = \frac{3!}{0!3!} = 1 \qquad (4\text{-}48)$$

Note that 0! is *defined* as 1. When we look at the group $x^2y$ and $xy^2$ from the point of view of Eq. (4-47), we have

$$\binom{N}{n} = \binom{3}{2} = \frac{3!}{1!2!} = 3 \qquad (4\text{-}49)$$

where 1! also equals 1. That this is correct is shown by the arrangements *xxy*, *xyx*, *yxx*, and *xyy*, *yxy*, *yyx*. When we say we are selecting three things two at a time, in this case we mean that we are taking a group of three numbers that can be distinguished as either one of two kinds, $x$ or $y$.

## Glossary

*Bayes' Theorem*. Theorem for assigning a quantitative change to the probability that a hypothesis is true given the result of a test (observation, experiment) of the hypothesis.

*Binomial Coefficient*. Premultiplying constant preceding each term of an expanded binomial series.

*Carrier*. Heterozygote.

*Combinations*. The number of distinct groupings of $N$ things taken $n$ at a time.

*Conditional Probability*. Probability of one event occurring after another event, on which it depends, is known to have occurred.

*Factorial*. The product obtained by multiplying a number by itself minus one, that product by the number minus two and so on down to one: $N! = N(N-1)(N-2)\ldots(1)$.

*Genotype*. A group or class sharing a specific genetic constitution, e.g., *XX* or *Xx*.

*Heterozygote*. Member of a genotype carrying both genes, dominant and recessive.

*Homozygote*. Member of a genotype carrying only one kind of gene, dominant or recessive.

*Independent Events*. Two or more events for which the outcome of no one has an influence on any of the others.

*Permutations*. The number of ways in which a set of things may be arranged.

*Probability*. The number of ways an event can occur divided by the number of ways all events can occur, provided all events are equally likely.

*Real Variable*. Strictly, any real number (as distinct from imaginary or complex), but used in computer programing to designate a variable that may take on nonintegral values.

## Problems

*1*. What is the probability of throwing HHH with three coins?

*2*. Write down all the possible events that can occur when four coins are thrown simultaneously. Verify that they total sixteen.

*3*. Rewrite Program 4-1 to accommodate a system that generates only decimals from 0 to 1. You might wish to use the INT($n$) statement, which truncates any number $n$ to its next lower integer. You will have to be careful not to bias the results to lower integers and you will have to exclude scores of 0.

*4*. Some systems do not accept IF. . .AND. . .THEN statements. Rewrite Program 4-2 using only IF. . .THEN statements.

*5*. Two individuals are diagnosed with a disease having a 60% mortality rate. What is the probability of 0, 1, 2 survivals?

*6*. What is the probability $p$((7 or 11) and T) for tossing two dice and a coin? What is the probability $p$(7 or (11 and T))?

*7*. What is the probability of drawing a four or a jack from a well shuffled deck of cards? What is the probability of drawing a four and a jack?

*8*. A certain disease is known to have a fatality rate of 18 patients per hundred diagnosed cases. What is the probability of diagnosing two successive cases of the disease both of which will prove to be fatal?

*9*. In how many ways can two books be chosen from five different books and arranged in two spaces on a shelf? How many pairings are possible without regard for arrangement?

*10*. A bag contains 20 blue balls and 30 white balls. What is the probability of drawing two successive blue balls out of this bag if the first ball is not replaced in the bag?

*11*. If 20% of the beads in a *large box* are colored green, what is the probability of getting 2 and only 2 green beads in a sample of 3? What is the probability of getting at least 2 green beads?

*12*. It is interesting to watch frequencies of simulated events converge on the theoretical probability as the number of trials is increased. Modify Program 4-1 so that it simulates only 10 rolls of one die and prints out the mean score on each iteration. Run the program several times. Observe the fluctuations of M about the theoretical value of 3.500. Is the theoretical value *ever* observed?

## Bibliography

W. S. Dorn, H. J. Greenberg, and M. K. Keller, *Mathematical Logic and Probability with BASIC Programming,* Prindle, Webber, and Schmidt, Boston, Mass., 1973.

S. L. Meyer, *Data Analysis for Scientists and Engineers,* Wiley, New York, N.Y., 1975.

M. R. Spiegel, *Theory and Problems of Statistics,* Schaum's Outline Series, McGraw-Hill, New York, N.Y., 1961.

H. D. Young, *Statistical Treatment of Experimental Data,* McGraw-Hill, New York, N.Y., 1962.

# Chapter 5

# Determining Probability Distributions

There are many situations in which only one of two possible outcomes may be observed. A coin may fall heads or tails; an organism may be infected by a virus or not infected by it; a gene may be dominant or recessive. If we observe $N$ events, there is a probability $p(n)$ that $n$ of them will turn out one way and $N - n$ will turn out the other way. There are also probabilities that the number of outcomes of one kind will not be $n$, but some other number. Taken together, all probabilities constitute a *distribution*. The distribution of $p(n)$ in which events can occur only one of two ways is the *binomial distribution*. This chapter introduces distributions using the binomial case (there are others) first for the situation that the probability of observing an event is the same as the probability of observing its opposite, as in the case of coin tossing. The chapter continues by describing cases in which the probabilities are different and concludes with applications in the biomedical field.

If we throw 8 coins simultaneously, there are 9 combinations of heads and tails (H,T) that can occur, 8H, 7H-T, 6H-2T, . . ., H-7T, 8T. These events can be distinguished by the number of heads $n$H and when we know $n$, we know that the number of tails is $9 - n$. For each value of $n$ there is a specific probability; that is, the probability that $n$ heads will be thrown is a function of $n$. This is written

$$p(n) = F(n) \tag{5-1}$$

read "$p$ of $n$ is a function of $n$." If all eventualities are considered,

$$\sum p(n) = 1 \tag{5-2}$$

All the values $p(n)$ taken together constitute a *probability distribution* and $p(n)$ is said to be a *probability function*.

## Tree Diagrams

Probability functions can be represented by *tree diagrams* and histograms. Tree diagrams are more appropriate for distributions over a small number of possible events, and histograms are more appropriate for large distributions. The tree diagram corresponding to the toss of one coin is given in Fig. 5-1 and the corresponding histogram in Fig. 5-2.

The toss of two coins simultaneously or sequentially has been discussed and is represented by the tree diagram in Fig. 5-3. In each case, the independent variable, plotted on the horizontal axis, is the number of "tails" observed. The value is 0 in the HH case, 1 in either the HT or TH case without distinction and 2 in the TT case.

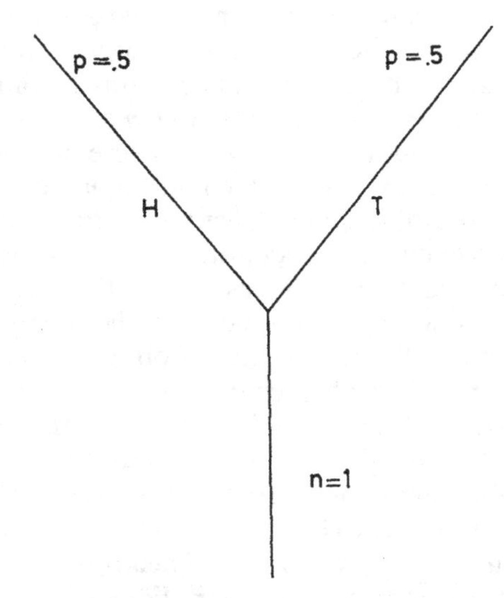


Fig. 5-1. Tree diagram for the toss of one coin.

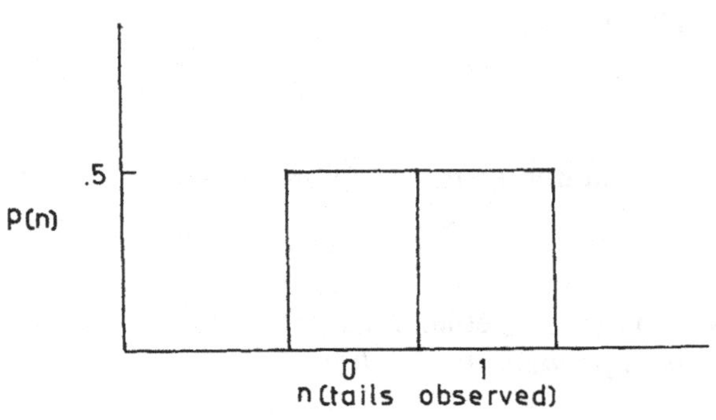


Fig. 5-2. Histogram for the toss of one coin.

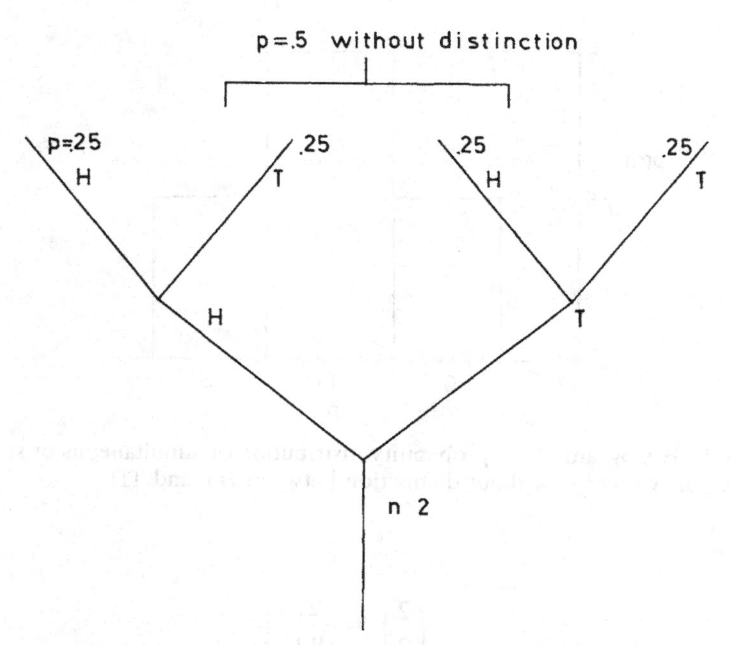


Fig. 5-3. Tree diagram for simultaneous or sequential tosses of two coins.

The ratio of probability functions can be calculated from the binomial coefficients. In the first case, $n = 0$ or 1 since we observe T either 0 times or 1 time.

$$\binom{N}{n} = \binom{1}{0} = \frac{N!}{(N-n)!n!} = \frac{1!}{1!0!} = 1$$

and

$$\binom{N}{n} = \binom{1}{1} = \frac{N!}{(N-n)!n!} = \frac{1!}{0!1!} = 1 \tag{5-3}$$

The ratio $p_H{:}p_T$ is 1:1. In the second, $N = 2$ (Fig. 5-4):

$$\binom{N}{n} = \binom{2}{0} = \frac{2!}{2!0!} = 1$$

$$\binom{2}{1} = \frac{2!}{1!1!} = 2$$

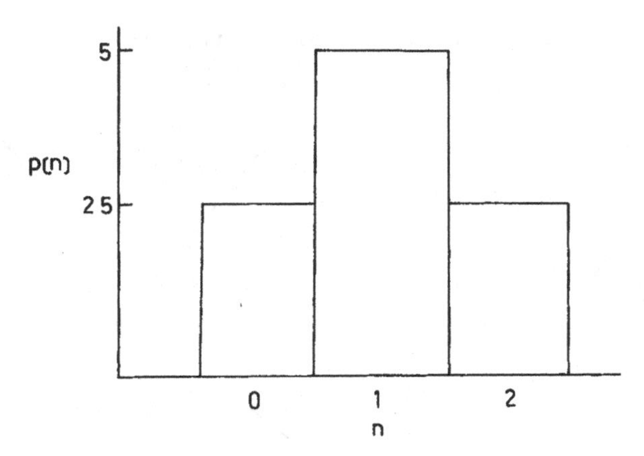


Fig. 5-4. Histogram of the probability distribution of simultaneous or sequential tosses of two coins without distinction between HT and TH.

and

$$\binom{2}{2} = \frac{2!}{0!2!} = 1 \tag{5-4}$$

The ratio $p_{HH}:p_{HT,TH}:p_{TT}$ is 1:2:1.

Notice that we can arrive at the absolute probabilities from their ratio by applying what is called a *normalization condition*. In the case of the toss of one coin, we can see by inspection that the two numbers that have the ratio 1:1 and the sum 1 are 1/2 and 1/2. In the two-coin case, summation of the non-normalized probabilities, 1, 2, 1, is 4. Dividing each by 4 leads to 1/4, 1/2, and 1/4 with $\Sigma p = 1$.

The tree diagram for three tosses of a coin is shown in Fig. 5-5. If no distinction is made between groups such as HTT and THT, the histogram for three tosses follows from the binomial coefficients, which give the ratio of the probabilities, and the normalization condition that $\Sigma p(n) = 1$ if all eventualities are considered. It was shown in the last chapter that the binomial coefficients of three independent events that can occur in either one of two ways are 1:3:3:1.

The absolute height of the columns in Fig. 5-6 can be calculated from the restrictions $3h_0 = h_1 = h_2$, $3h_3 = h_2$ and $h_0 + h_1 + h_2 + h_3 = 1$. The solution is an exercise in simultaneous equations for a very restricted set. General methods for solutions of simultaneous equations and computer solution will be described in Chapter 11. Here, if $h_1 = h_2$, the equation of restraint, $h_0 + h_1 + h_2 + h_3 = 1$, can be simplified to

$$h_0 + 2h_1 + h_3 = 1 \tag{5-5}$$

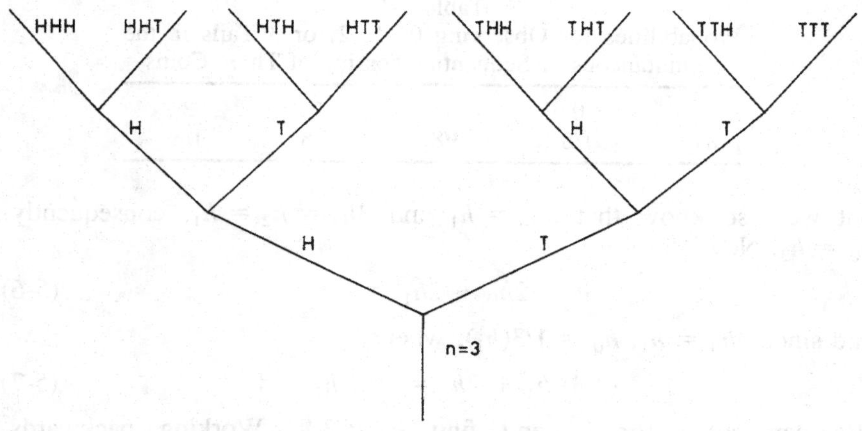


Fig. 5-5. Tree diagram for three simultaneous or sequential tosses of a coin.

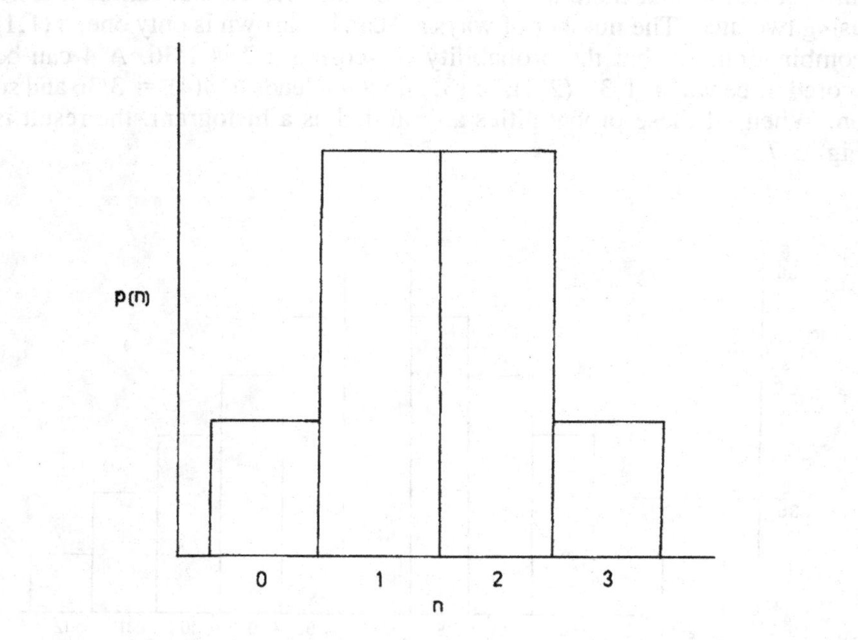


Fig. 5-6. Histogram for three simultaneous or sequential tosses of a coin.

Table 5-1
Probabilities for Observing 0, 1, 2, or 3 Tails in the Simultaneous or Sequential Tossing of Three Coins

| $n$ | 0 | 1 | 2 | 3 |
|---|---|---|---|---|
| $p(n)$ | 1/8 | 3/8 | 3/8 | 1/8 |

but we also know that $3h_0 = h_1$ and $3h_3 = h_2 = h_1$, consequently, $h_0 = h_3$. Now,

$$2h_0 + 2h_1 = 1 \tag{5-6}$$

and since $3h_0 = h_1$, $h_0 = 1/3(h_1)$, whence

$$(2/3)\ h_1 + 2h_1 = (8/3)h_1 = 1 \tag{5-7}$$

We can solve for $h_1$ and find $h_1 = 3/8$. Working backwards, $h_2 = h_1 = 3/8$ and $h_0 = h_3 = 1/3\ h_1 = 1/8$. We are now able to determine specific values of the probabilities as shown in Table 5-1. The summation of values in Tables 5-1 is one, as it must be.

The probability distribution of $p(n)$ vs $n$ for throwing two dice can be built up by consideration of the number of ways each number ($n$) can be scored, divided by the number of ways all numbers can be scored. We have observed that there are $6^2 = 36$ possible scores that can be thrown using two dice. The number of ways a 2 can be thrown is only one, a (1,1) combination, so that the probability of scoring a 2 is 1/36. A 4 can be scored three ways (1,3), (2,2), or (3,1), which leads to $p(4) = 3/36$ and so on. When all these probabilities are plotted as a histogram, the result is Fig. 5-7.

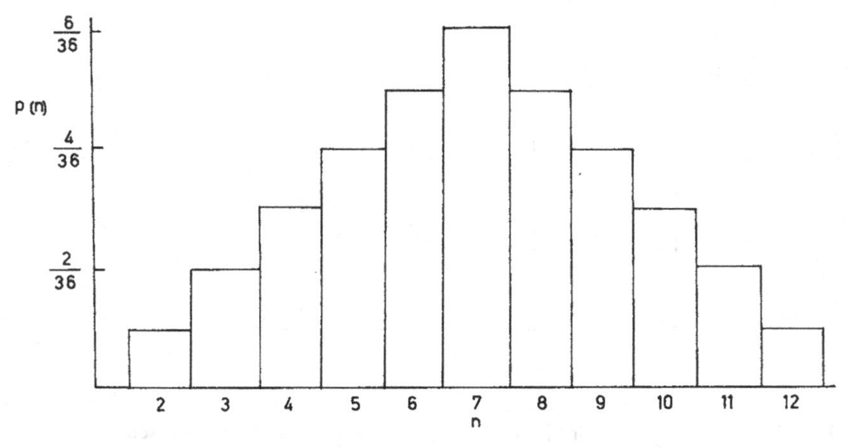


Fig. 5-7. Histogram of the probability function for throwing two dice.

## Application: Heredity

Suppose we wish to breed the recessive trait *xx* out of a population of animals. The trait is uniformly distributed among the animals such that, having removed all *xx* genotypes, we have a randomly distributed proportion of heterozygous carriers. We wish to decide which inbreeding technique, mother–son mating or brother–sister mating will be more effective in reducing the *xx* genotype.

Qualitatively, we may answer the question in the following way. By preventing *xx* genotypes from breeding, the strain gradually loses the *xx* recessive trait. After many generations, occurence of the *xx* trait will become vanishingly small. Since there is no reason to suppose any kind of nonrandom fluctuation in occurence of the recessive trait, we conclude that it disappears gradually and monotonically, a little less being observed in each generation. If this is so, we are led to conclude that any genetic pairing of two individuals from the second generation is more favorable to the rapid disappearance of the *xx* genotype than genetic pairing of an individual from the first generation with one from the second. To answer the problem specifically, the sister–brother pairing is more effective than the mother–son pairing.

Perhaps we do not agree with the preceeding qualitative logic or perhaps we wish to have quantitative knowledge of how rapidly the recessive genotype diminishes using either inbreeding technique. Probability equations must be developed from a quantitative model. The model is best set up by constructing two tree diagrams indicating the various probabilities. Let the proportion of heterozygous members of the population be $\alpha$, whence the proportion of homozygous *XX* genotypes is $(1 - \alpha)$.

Suppose a male is selected from the population for breeding. The probability that he is an *Xx* genotype is $\alpha$ and the probability that he is *XX* genotype is $(1 - \alpha)$. A female is selected for breeding and the same probabilities apply. Selection of the male, M, is shown by the first branching of the tree diagram in Fig. 5-8, and selection of the female, F, is shown by the next higher branchings. The four possible genetic pairings are shown by the four circles above the second branching, from left to right, M, F both *XX*; M *XX*, F *Xx*; M*Xx*, F *XX*; M*Xx*, F *Xx*.

Genotypes of the male offspring of the four possible pairings are shown in the top row of eight circles. We are interested only in the male offspring at this point because we are determining the genotypes of prospective mother–son mating pairs; the genotypes of the mothers and the probabilities of their occurrance are, of course, already known.

The probability of observing each genotype as a result of each genetic pairing is shown above the appropriate circle, each in terms of some multiple of $\alpha$, the proportion of heterozygotes in the original population. The probability of occurance of each genotype depends on the probabilities of

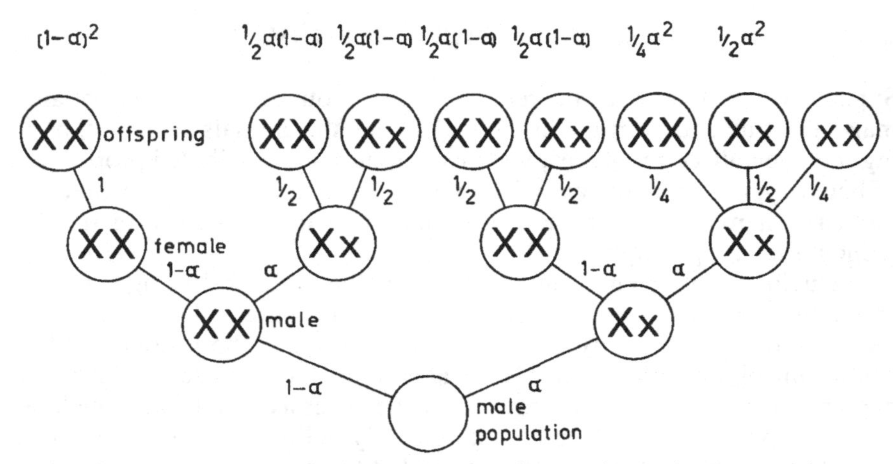


Fig. 5-8. Tree diagram for transmission of genes in mother–son mating.

three sequential but independent events: (1) selection of the male parent, (2) selection of the female parent, and (3) random combination of the parent's genes to determine the genotype of the offspring. Reading from left to right in Fig. 5-8, selection of the male *XX* homozygote has a probability of $(1 - \alpha)$, selection of the female homozygote has a probability of $(1 - \alpha)$ and, these events having occured, the probability of an *XX* homozygous offspring is 1. The product of these three independent probabilities is $(1 - \alpha)^2$ as shown. In the next two cases, mating of an *XX* male with a *Xx* female produces homozygous and heterozygous offspring with equal probability. The next two cases are mathematically identical to the previous two, except that the genotype of the male and female are reversed. The last three cases on the right represent the probabilities of homozygous and heterozygous genotypes discussed in the previous chapter resulting from heterozygous parents. The last case on the right, the *xx* recessive genotype has no number above it (the calculated number is $1/4\alpha^2$) because the *xx* genotype will not be used for breeding and the probability associated with its appearance does not figure in the calculations below.

Of the eight mother–son pairs, seven are possible breeding pairs, one having been excluded by its physically distinguishable recessive trait. The other seven are, however, physically indistinguishable. Of these, two (numbers 3 and 7 from the left) combine a male and female that are both carriers, while the remaining five have at least one *XX* genotype. The probability that both mother and son are carriers, $p_c$, is the sum of the individual probabilities for cases 3 and 7

$$p_c = (1/2)\alpha(1 - \alpha) + (1/2)\alpha^2 = (1/2)\alpha \qquad (5\text{-}8)$$

The probability that either the mother or the son (or both) is a homozygote, $p_h$, is $1 - 1/2\alpha - 1/4\alpha^2$.

Both of these probabilities are relative to all genotypes that may be born. These are not the probabilities we seek however, because we have excluded the *xx* genotype. We wish to have the probability that both mother and son are carriers within the breeding population exclusive of the *xx* genotype. The probability that mother and son are not *xx* is $p_c + p_h$; hence the carrier probability relative to the breeding population is

$$p_1 = \frac{p_c}{p_c + p_h} \quad (5\text{-}9)$$

where $p_1$ refers to the probability of carrier pairs by the first (mother–son) mating scheme. Now,

$$p_1 = \frac{(1/2)\alpha}{1 - (1/2)\alpha - (1/4)\alpha^2 + (1/2)\alpha} = \frac{(1/2)\alpha}{1 - (1/4)\alpha^2} \quad (5\text{-}10)$$

**Exercise 5-1.**

Prove that

$$p_h = 1 - (1/2)\alpha + (1/4)\alpha^2$$

for the mother–son mating scheme diagrammed in Fig. 5-8.

The tree diagram for mating scheme 2, sister–brother mating, is shown in Fig. 5-9. It is identical to the previous one for the first three rows, diagramming the possible genotypes for father, mother, and son. The added top row indicates the possible pairings from each of the mother–father pairs. Note that, in each case, the number of male genotypes is the same as the number of female genotypes.

The last statement is exemplified in the brother–sister pair at the extreme left of Fig. 5-9. Since both mother and father are homozygous *XX*, both brother and sister are homozygous *XX*, and only that mating combination is possible. Since the genotype of both brother and sister depends upon simultaneous selection of *XX* parents, the probability of simultaneous independent selection of parents of this genotype is the product $(1 - \alpha)$ $(1 - \alpha)$. Note that this probability is relative to the total number of genotypes that would be produced if no scheme of selective breeding were followed. The second mother–father pair can produce two genotypes, *XX* and *Xx*; hence either son or daughter can be homozygous *XX* or a carrier. The total number of mating combinations is four, with genotypes shown in circles and probabilities given by the formulas above the circles 2 through 5 at the top. The next four circles are identical to the four preceding them because one parent is a carrier but the other is not.

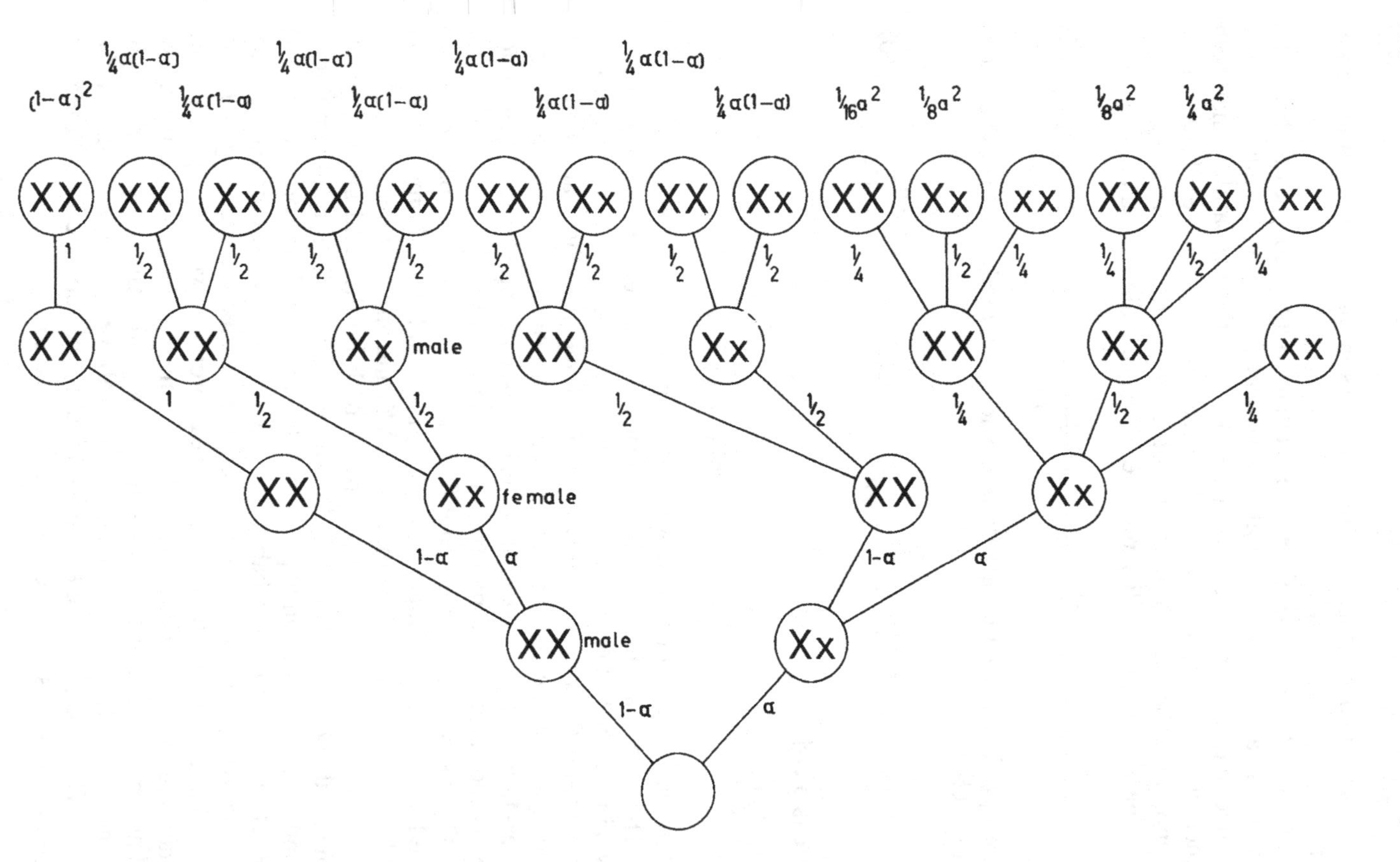


Fig. 5-9. Tree diagram for gene transmission for sister–brother mating.

When both parents are carriers, all three genotypes are possible for both son and daughter. Thus the right side of Fig. 5-9 shows three possible genetic pairings with the *XX* son and three with the *Xx* son. No pairings are indicated for the *xx* recessive male genotype and no probabilities are written above the *xx* recessive female genotype since these are not bred. Probabilities appear above the tree diagram only for those pairings that are selected by the absence of the observable *xx* recessive trait in both mates.

We obtain the probability that both mates of a pair are carriers in exactly the same way that we did in the simpler mother–son mating scheme. First the probability that both are carriers relative to all possible genotypes is calculated simply by summing the probabilities written above any *Xx*, *Xx* pair.

$$p_c = (1/2)\,\alpha - (1/4)\,\alpha^2 \qquad (5\text{-}11)$$

where the appropriate mating pairs are 5, 9, and 14 taken from left to right.

Similarly, the probability that at least one mate is homozygous *XX* is the sum of the remaining probabilities

$$p_h = 1 - (1/2)\,\alpha - (3/16)\,\alpha^2 \qquad (5\text{-}12)$$

Now, to convert from probabilities relative to all possible genotypes to probabilities relative to genotypes selected for breeding, we take

$$p_2 = \frac{(1/2)\,\alpha - (1/4)\,\alpha^2}{1 - (7/16)\,\alpha^2} \qquad (5\text{-}13)$$

where $p_2$ is the probability that a mating pair will both be carriers in the second (sister–brother) breeding scheme.

We wish to determine the effectiveness of each scheme in reducing the *xx* recessive trait in the population and, quantitatively, how the proportion, $\alpha$, of carriers in the original population affects $p_1$ and $p_2$, the expected carrier population in the third generation obtained by either mother–son or brother–sister mating.

## Exercise 5-2

Prove that

$$p_c = (1/2)\,\alpha - (1/4)\,\alpha^2$$

for the sister–brother mating scheme diagrammed in Fig. 5-9.

## Exercise 5-3

Prove Eqs. (5-12) and (5-13).

## Program 5-1

Reviewing the original problem: we wish to know whether scheme 1 or scheme 2 is best for breeding out the *xx* trait. If $p_1$ is greater than $p_2$, there will be more carriers produced by scheme 1 and if $p_2 > p_1$ there will be more carriers produced by scheme 2. We shall write a computer program to determine $p_1$ and $p_2$, for several values of $\alpha$, plot the results and so obtain a graphical expression of our conclusions. We might also ask whether $p_1$ might be the more favorable breeding scheme for some values of $\alpha$ and $p_2$ more favorable for others. Graphical representation of the computer output will provide clear answers to these questions.

```
5    PRINT'      A                P1               P2'
6    PRINT' -------          -------          -------'
10   FOR A=0.05 TO 1 STEP 0.05
20   P1=0.5*A/(1-0.25*A*A)
21   P2=(0.5*A-0.25*A*A)/(1-0.4375*A*A)
30   PRINTUSING 40, A, P1,P2
40   :##.#####       ##.#####       ##.#####
50   NEXT A
60   END

READY
RUNNH

     A              P1              P2
  -------        -------         -------
  0.05000        0.02502         0.02440
  0.10000        0.05013         0.04771
  0.15000        0.07542         0.07006
  0.20000        0.10101         0.09160
  0.25000        0.12698         0.11245
  0.30000        0.15345         0.13273
  0.35000        0.18053         0.15255
  0.40000        0.20833         0.17204
  0.45000        0.23700         0.19133
  0.50000        0.26667         0.21053
  0.55000        0.29750         0.22979
  0.60000        0.32967         0.24926
  0.65000        0.36338         0.26912
  0.70000        0.39886         0.28958
  0.75000        0.43636         0.31088
  0.80000        0.47619         0.33333
  0.85000        0.51869         0.35732
  0.90000        0.56426         0.38335
  0.95000        0.61340         0.41208
  1.00000        0.66667         0.44444

TIME:   0.94 SECS.
```

**Commentary on Program 5-1.** Statements 5 and 6 provide the headings for a table of alpha, $p_1$ and $p_2$ which, to conform with teletype printing capabilities, are called A, P1, and P2. The calculations of probabilities P1 and P2 are contained in a FOR–NEXT loop that functions much as

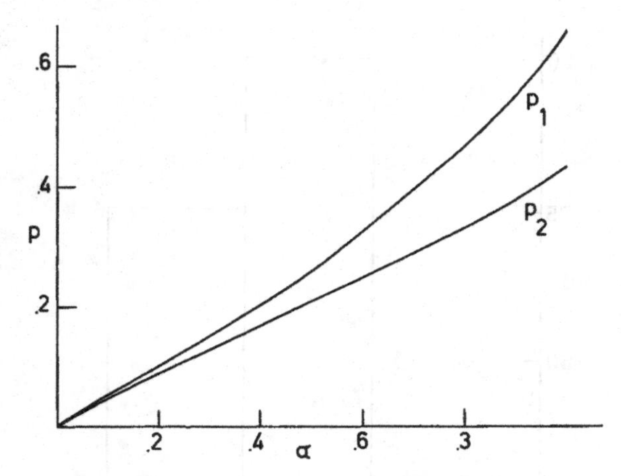


Fig. 5-10. Expected carrier population of offspring by breeding schemes 1 and 2.

those already encountered, except that the interval is smaller and the steps are specified as 0.05 in units of A. This information is introduced in statement 10, which is an ordinary FOR statement with a step size of 0.05. The equations to be solved in each step are given in statements 20 and 21.

The PRINT module is not different from those already encountered in Program 4-1. The PRINTUSING statement with its necessary format symbols determines the number of digits on either side of the decimal point, either none or one, as comparison with the printout shows.

The graphical results shown in Fig. 5-10 indicate that the brother–sister mating scheme, scheme 2, is always favorable to elimination of the recessive trait, no matter what the value of alpha.

## Modal and Mean Scores

Suppose we make up a game in which two coins are tossed and a score of 1 is awarded for HH, 2 is awarded for HT or TH, and 3 is awarded for TT. The modal toss is HT or TH, as already shown.

The score of 1 for HH will be recorded with a relative frequency of 0.25 for many tosses. The score of 2 will be recorded with a frequency of 0.50, and the score of 3 with a frequency of 0.25. The anticipated mean score is $0.25(1) + 0.50(2) + 0.25(3) = 1/4 + 4/4 + 3/4 = 2$. The histogram of scores is shown in Figure 5-11.

In general, if we assign any set of numbers to a set of events and observe many events, say $Z$ events ($Z$ is large), the mean score $\bar{n}$ is

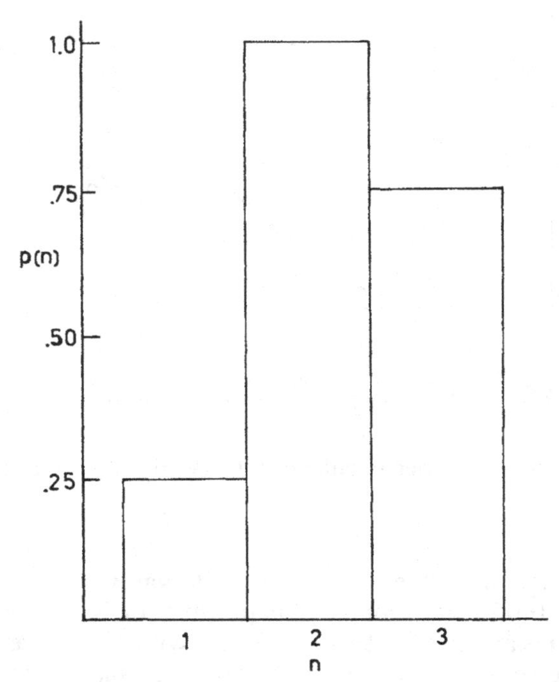


Fig. 5-11. Histogram of scores for throwing two coins with HH scoring 1, HT or TH scoring 2, and TT scoring 3.

$$\bar{n} = \frac{1}{Z}\sum nZp(n) = \sum np(n) \tag{5-14}$$

as we saw in the coin-tossing example above.

Note that in this and similar calculations the probability that event $n$ will occur, $p(n)$, is the anticipated relative frequency of future events, $f(n)$, provided that enough trials are made. Probabilities may be calculated from the mechanical nature of the system as they are for coins that have only two equally probable ways they can turn up, $p(n) = 1/2$, and dice which have six, $p(n) = 1/6$. Frequently we do not know the mechanics of the system we are studying and cannot make an *a priori* calculation. A case in point would be the incidence of infection of plants or animals exposed to a certain virus. We do not have detailed information of the mechanism of infection by viruses; hence we take the frequency of past infection relative to exposure as the probability to be used in predicting the number of infections that will result from future exposure of a population. We are using the common-sense notion that future events can be predicted from the summa-

Table 5-2
Probabilities of Throws and of Scores for Two Dice

| $n$ | $p(n)$ | $np(n)$ | $n$ | $p(n)$ | $np(n)$ |
|---|---|---|---|---|---|
| 2 | 1/36 | 2/36 | 8 | 5/36 | 40/36 |
| 3 | 2/36 | 6/36 | 9 | 4/36 | 36/36 |
| 4 | 3/36 | 12/36 | 10 | 3/36 | 30/36 |
| 5 | 4/36 | 20/36 | 11 | 2/36 | 22/36 |
| 6 | 5/36 | 30/36 | 12 | 1/36 | 12/36 |
| 7 | 6/36 | 42/36 | | | |

tion of our past experience. It might be objected that we may make an error in this way and the objection is quite valid. Statistics and probability are used when there is no direct way of calculating a result, and we wish to make the best possible educated guess. The chance of statistical error is more widespread in the physical sciences than is generally appreciated. Even the laws of thermodynamics are statistical in nature, and there is a minute but finite probability that they will fail in any specific situation.

A more complicated situation than that of throwing two coins is that of throwing two dice, computing the expected mean score, and drawing the histogram of expected scores. Probable throws and probable scores for two dice are shown in Table 5-2. The probabilities of throws ($n$) are given by $p(n)$ and the probabilities of scores by $np(n)$.

The mean value of scores for throwing two dice is expected to be

$$\bar{n} = \sum np(n) = 252/36 = 7 \tag{5-15}$$

The histogram of scores is given in Fig. 5-12. Note that the modal score is the same as the mean score although the probability function is not symmetrical. This is because a low roll is just as likely as a high roll and the deviations cancel.

The variance is the arithmetic mean of the sum of the squares of the deviations; hence, by analogy to Eq. (5-14)

$$\sigma^2 = \sum (n - \bar{n})^2 p(n) \tag{5-16}$$

Table 5-3 shows the deviations from $n$ of each throw, $p(n)$ and $(n - \bar{n})^2 p(n)$ for all possibilities. Application of Eq. (5-16) leads to a variance and standard deviation of

$$\sigma^2 = 210/36 = 5.83 \tag{5-17}$$

and

$$\sigma = \sqrt{5.83} = 2.42 \tag{5-18}$$

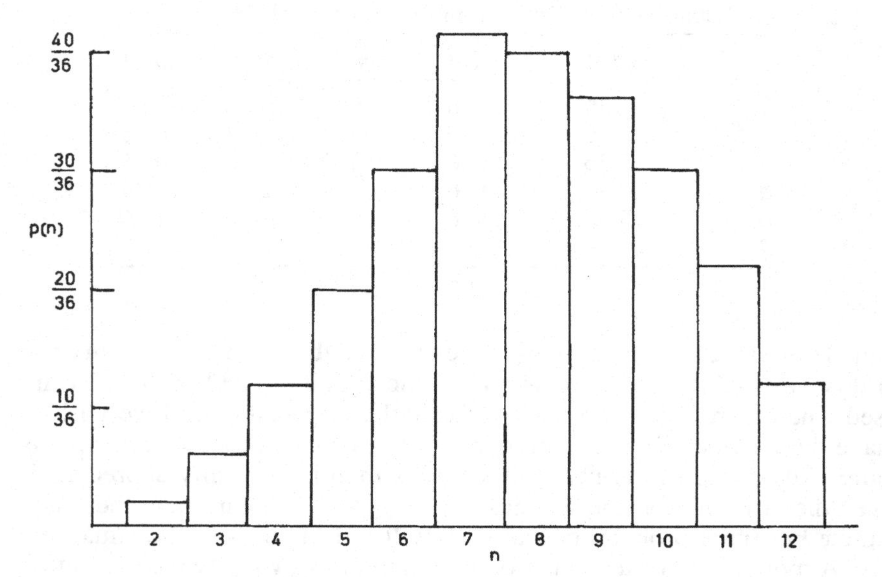


Fig. 5-12. Histogram of $n = \Sigma\, np(n)$ for two dice.

Table 5-3
Probable Deviations for the Binomial Distribution of Throwing Two Dice

| $n$ | $n - \bar{n}$ | $p(n)$ | $(n - \bar{n})^2 p(n)$ |
|---|---|---|---|
| 2 | −5 | 1/36 | 25/36 |
| 3 | −4 | 2/36 | 32/36 |
| 4 | −3 | 3/36 | 27/36 |
| 5 | −2 | 4/36 | 16/36 |
| 6 | −1 | 5/36 | 5/36 |
| 7 | 0 | 6/36 | 0/36 |
| 8 | 1 | 5/36 | 5/36 |
| 9 | 2 | 4/36 | 16/36 |
| 10 | 3 | 3/36 | 27/36 |
| 11 | 4 | 2/36 | 32/36 |
| 12 | 5 | 1/36 | 25/36 |

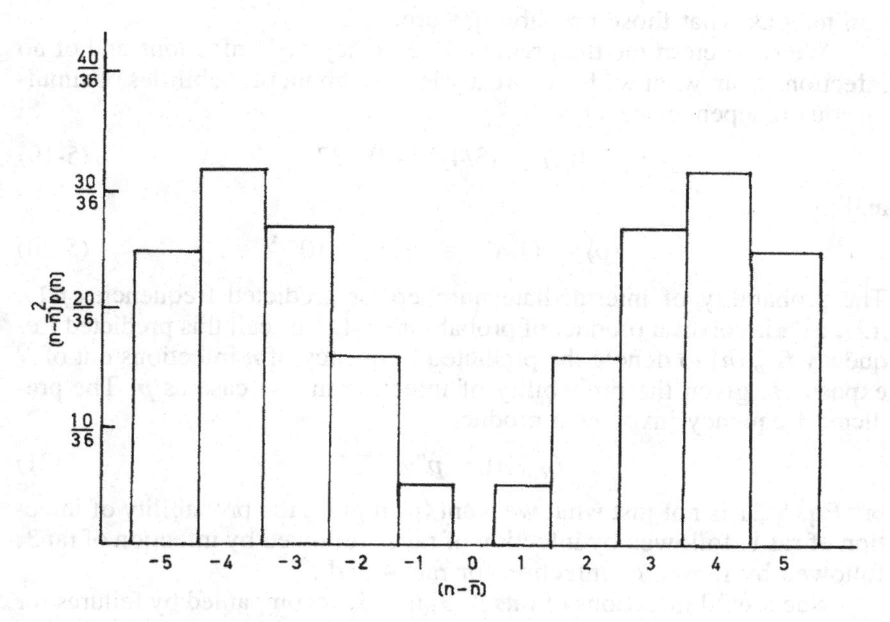


Fig. 5-13. Histogram for the squares of expected deviations for throwing two dice.

## The Binomial Distribution

We have studied binomial distributions by means of tree diagrams and histograms for symmetrical distributions. We shall now develop a powerful and general mathematical formula for treating any binomial distribution.

Suppose we wish to infect 5 experimental rats with a virus but, based on many previous experiments, we know that the probability of successful incubation of the virus is 3/4 and that the probability of no infection is 1/4. The infection of each rat is taken to be independent of the others; hence we are dealing with *N* independent events. They are binomially distributed because each rat either is or is not infected; there is no other possibility. The probability of successful incubation is $p$ and that of failure is $1 - p$, which is usually denoted $q$. For a binomial distribution, $p + q = 1$.

Since we are dealing with probabilities, not certainties, we cannot say that 3/4 of the rats will be infected. It would be absurd to say that 3/4 (5) = 3 ¾ rats will be infected while 1 ¼ remain healthy. There are finite probabilities that 0, 1, 2, 3, 4, or 5 rats will be infected and the binomial distribution tells us what those probabilities are.

We can determine the predicted frequency of 5 infections and of no infections from what we have already learned about probabilities of simultaneous independent events

$$f(5) = (3/4)^5 = 0.237 \tag{5-19}$$

and

$$f(0) = (1/4)^5 = 9.76 \times 10^{-4} \tag{5-20}$$

The probability of intermediate numbers or predicted frequencies $f(1)$, $f(2)$, . . . involves a product of probabilities. Let us call this predicted frequency $f_{N,p}\,(n)$ to denote the predicted frequency of $n$ infections out of $N$ exposures, given the probability of infection in any case as $p$. The predicted frequency involves a product

$$f_{N,p}(n) \sim p^n q^{N-n} \tag{5-21}$$

but Eq. 5-21 is not just what we want for it gives the probability of infection of rat 1, followed by infection of rat 2, followed by infection of rat 3, followed by failure of infection for rats 4 and 5.

Successful infections of rats 2, 3, and 5, accompanied by failures for rats 1 and 4, would be equally acceptable, as would a number of other combinations. The number of different ways of achieving success must be multiplied into the probability that each way will occur. This is the number of combinations of $N$ things taken $n$ at a time. Multiplication yields

$$f_{N,p} = \binom{N}{n} p^n q^{N-n} \tag{5-22}$$

which is called the binomial distribution, $f_{N,p}(n)$.
Now, let $n = 1$ and $N = 5$

$$\binom{N}{n} = \frac{N!}{(N-n)!n!} = \frac{120}{24(1)} = 5$$

$$p^n = (3/4)^1 = 0.750$$

$$q^{N-n} = (1/4)^4 = 0.00390$$

$$f_{N,p}(n) = 0.0146 \tag{5-23}$$

for the probability that any one, but only one, rat will be infected.

Let $n = 2$ and $N = 5$, whence

$$f_{N,p}(n) = 0.0879 \tag{5-24}$$

for the probability that any two but exactly two rats will be infected.

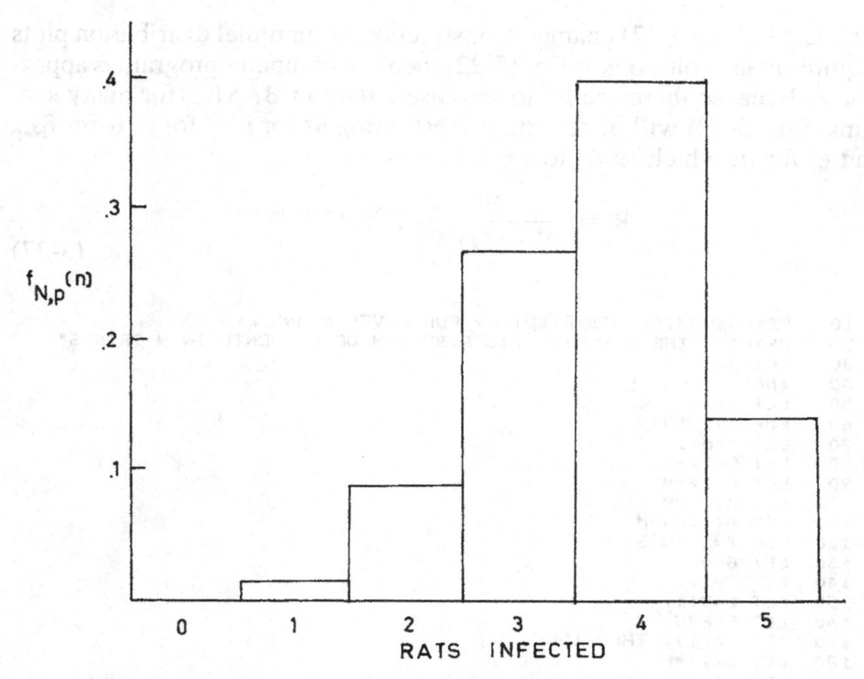


Fig. 5-14. Binomial distribution curve for $N = 5$, $p = 3/4$.

For $n = 3$ and $N = 5$,

$$f_{N,p}(n) = 0.2636 \tag{5-25}$$

and for $n = 4$ and $N = 5$

$$f_{N,p}(n) = 0.395 \tag{5-26}$$

The binomial histogram obtained from these results is shown in Fig. 5-14. It is unsymmetrical or skewed to the left because $p \neq q$.

## Exercise 5-3

Prove that

$$\sum f_{N,p}(n) = 1$$

for the binomial distribution.

## Program 5-2

Calculations of the kind carried out in Eqs. (5-23) to (5-26) are frequently required and can be tedious for a large number of trials. Moreover, it is interesting to see how the binomial distribution function changes as the parameters in Eq. (5-22) change. Construction of binomial distribution plots requires many solutions to Eq. (5-22); hence a computer program is appropriate. Because there are no lower case letters in BASIC (for many systems) Eq. (5-22) will be rewritten substituting $M$ for $n$, $P$ for $p$, R for $f_{N,p}$ and $Q$ for $q$, which leads to

$$\mathrm{R} = \frac{N!}{(N - M)!M!} P^M Q^{(N - M)} \tag{5-27}$$

```
10    REM BINOMIAL DISTRIBUTION FOR GIVEN N AND P
20    PRINT'   THE BINOMIAL DISTRIBUTION OF M EVENTS IN N TRIALS'
30    LET M=-1
40    INPUT N,P
50    LET J=N+1
60    FOR K=1 TO J
70    LET M=M+1
80    LET Q=1.-P
90    LET A(1)=N
100   LET A(2)=M
110   LET A(3)=N-M
120   FOR I=1 TO 3
130   LET G=0.
140   LET F=1.
150   LET G=G+1.
160   LET F=F*G
170   IF G<A(I), THEN 150
180   LET B(I)=F
190   NEXT I
200   LET C=B(1)/(B(2)*B(3))
210   LET R=C*P**M*Q**(N-M)
220   PRINT M,R
250   NEXT K
260   END

READY
```

**Commentary on Program 5-2.** After a remark and a title, $M$ is initialized at $-1$ and $n$ and $p$ are read in via an INPUT statement. The value of $M$ is initialized at $-1$ because it is to be incremented by 1 in the FOR–NEXT loop starting at 60, and we wish the first value of $M$ that enters the computation to be zero. We are now ready to enter the outer loop. The previous discussion shows us that there will be $N + 1$ computations for each binomial distribution, six in this case; hence J is set at $N + 1$ prior to entering the outer loop, which will be iterated up to J.

Upon entering the outer loop, $M$ is incremented to zero and $1 - P$ is stored at location Q. Program steps 70 through 200 repeat Program 4-2. At the end of this sequence of steps, we have the binomial coefficient, one on the first iteration, stored at location C. The next program step multiplies the binomial coefficient into $P^M Q^{(N-M)}$ to obtain the probability, $R$, that zero rats will be infected, assuming that the binomial distribution is followed. The value of $M$ and the resultant probability are printed out as the first line of output. The NEXT K statement sends control back to statement 60 to enter an iteration identical to the one just described, except that $M$ has now been incremented to one. Iterations are repeated with a printout on each loop until $M = 5$.

### Exercise 5-4

What are the probabilities that, in a family of four there will be 0, 1, 2, 3, or 4 girls? What is the most likely number of girls? Draw the appropriate histogram of $f_{N,p}(n)$ vs $n$. Take $p(\text{girls}) = 0.5$.

Solution 5-4. Computer output for this exercise is

```
THE BINOMIAL DISTRIBUTION OF M EVENTS IN N TRIALS
?4,.5
0                   0.0625
1                   0.25
2                   0.375
3                   0.25
4                   0.0625


TIME:  0.42 SECS.
```

### Exercise 5-5

Run Program 5-2. Obtain the binomial distribution for $N = 10$ and $p = 0.1$, 0.5, and 0.75.

## Effect of $p$ on the Binomial Distribution

The results of Exercise 5-5 can be plotted to obtain the histogram in Fig. 5-15. The value $N = 10$ was selected arbitrarily, but the result is generally true: if $p$ is less than 0.5, the distribution lies predominantly in the area with $n < N/2$; if $p = 0.5$, the distribution is symmetrical about a maximum at $n = N/2$; and if $p > 0.5$, the distribution lies predominantly in the area having $n > N/2$. If $n$ is plotted starting at zero and increasing from left to right, as it normally is, distributions with $p < 0.5$ lie to the left and $p > 0.5$ to the right of the symmetrical case. The degree of displacement from $n = N/2$ and the asymmetry of the histogram are determined by $p$; both are large if $p$ is very different from 0.5 and small if it differs only slightly. These effects are also illustrated by Fig. 5-15.

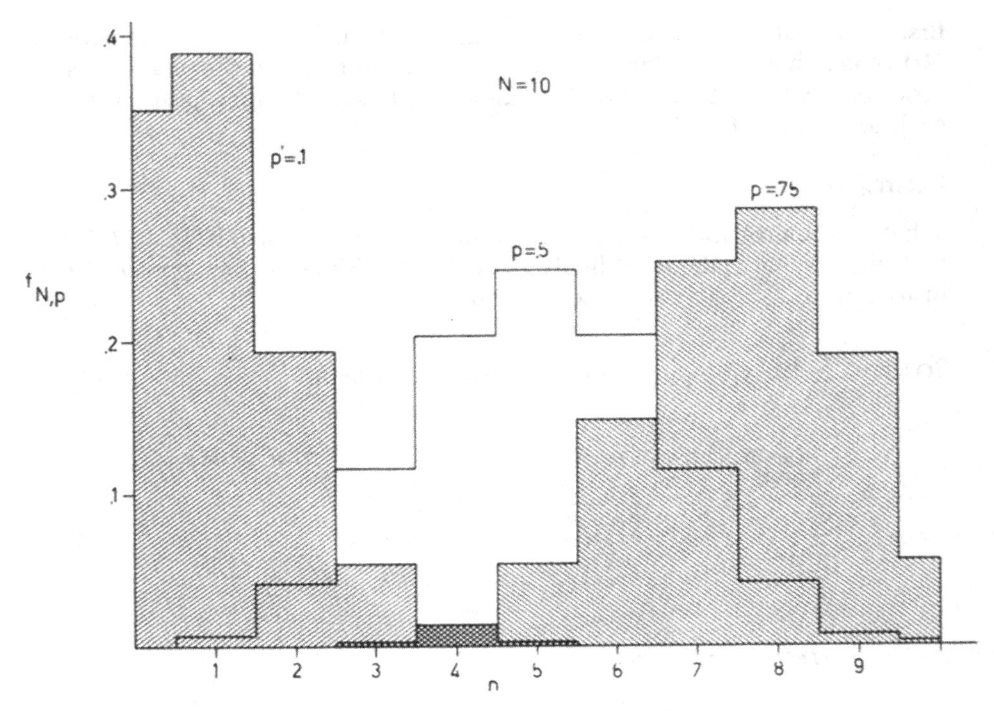


Fig. 5-15. Effect of $p$ on the binomial distribution.

## Mean Occurrences for Data Groups of *N*

If we make $N$ observations of success $n$ or failure $N - n$ of an event governed by the probability, $p$, then repeat the entire process $Z$ times where $Z$ is large, the mean number of successes $\bar{n}$ is

$$\bar{n} = \sum n/Z \tag{5-28}$$

By the same line of reasoning that lead to Eq. (5-14), we have

$$\bar{n} = \sum_{0}^{N} n \frac{N!}{(N - n)!n!} p^n q^{N-n} \tag{5-29}$$

It can be shown that $\bar{n} = Np$, which is the most convenient way of calculating the number of successes. This will be illustrated by considering the problem of a vaccination that has a 0.8 probability of success; hence a 0.2 probability of not "taking," or failure. If three individuals are vaccinated, we would like to know the probability that there will be 0, 1, 2, or 3 *failures* in the vaccination procedure. (Note that the form of the output data

Table 5-4
Predicted Frequencies of Vaccination Failure for Three Individuals ($q = 0.2$)

| $n$ | $f_{N,q(n)}$ | $nf_{N,q(n)}$ |
|---|---|---|
| M = 0 | 0.5119 | 0.0000 |
| M = 1 | 0.3839 | 0.3839 |
| M = 2 | 0.0959 | 0.1918 |
| M = 3 | 0.0079 | 0.0237 |

depends upon the way we phrase our question, in terms of success or failure.) Calculation of $f_{3,0.2}$ and $nf_{3,0.2}$ leads to the data in Table 5-4.

The most probable frequency of failure is zero for this vaccine. The mean result is not the same as the most probable result for this data set which is skewed because $p \neq q$. A little thought shows that the mean result cannot be equal to the most probable result for any vaccine having a good, but not perfect, probability of vaccination success. The probability of vaccination success being good, the most probable number of vaccination failures in a small number of trials is zero. The vaccine is not perfect, however; hence there must be some failures. The sum of these failures divided by $Z$, the number of sets of experimental results cannot be zero, no matter how large $Z$ may be.

The mean result, $\bar{n}$, can be calculated from Eq. 5-29 or from $Nq = 0.6$ (for failures). The mean number of successes is $Np = 2.4$ and the most probable number of successes is 3.

The variance is $\overline{(n - n)^2}$ where

$$\overline{(n - \bar{n})^2} = \sum(n - \bar{n})^2/Z = \sum(n - Np)^2 f_{N,p}(n) \qquad (5\text{-}30)$$

by the reasoning that led to Eq. 5-16.

This can be simplified (see problems at the end of this chapter) to

$$\sigma^2 = Npq \qquad (5\text{-}31)$$

or

$$\sigma = (Npq)^{1/2} \qquad (5\text{-}32)$$

so that we can calculate the variance and standard deviation for our vaccination problem as

$$\sigma^2 = 3(0.8)(0.2) = 0.48 \qquad (5\text{-}33)$$

and

$$\sigma = \sqrt{0.48} = 0.70 \qquad (5\text{-}34)$$

## Application: Evaluation of a New Treatment

Suppose a new treatment is implemented for a disease that, over many years, has been shown to be fatal in 60% of diagnosed cases. After the new treatment has been instituted, the first patient with the disease survives. We would like to know the probability that this survival owes to chance factors alone. Simple probability leads to the difference between one and the fractional fatality rate, $1.0 - 0.60 = 0.40$ or 40%. One survival gives us little information on which to base any evaluation of the new treatment.

If, however, a second patient given the same treatment survives, we begin to feel intuitively that there may be some merit in the new procedure. The probability of two chance survivals in two cases is $(0.40)^2 = 0.16$ or 16%. The probabilities of three, four, and five chance survivals are 0.064, 0.0256, and 0.01024, respectively. As the probability of observing a number of survivals in the new treatment owing merely to chance decreases, our confidence in the new treatment increases. In the latter two cases, we have a good indication that survival results from the improved treatment, although, if we adopt that point of view, there is still some chance that we will be wrong.

It may be, however, that not all patients survive, whether the new treatment is an improvement over the old one or not. If four patients survive out of five diagnosed cases, we must take the probability of chance survivals in *at least* four cases, i.e., the probability of four survivals plus the probability of five survivals. We already know that the latter probability is 0.0102 and the probability of exactly four survivals is calculated from the binomial formula

$$p = \frac{N!}{(N - n)\,!n!} p^n q^{N - n}$$

$$= 5(0.40)^4(0.60)^1 = 5(0.0256)(0.60)$$

$$= 0.0768$$

or 7.68%. The probability of at least four survivals is $7.68 + 1.02 = 8.70\%$ by chance alone. Hence observation of four survivals in the first five diagnosed cases is evidence in favor of the new treatment, but it is not as strong as four survivals out of four cases, as comparison with the previous paragraph shows.

## Glossary

*Binomial Distribution*. Probability distribution for events that can occur in only two ways.

*Carrier*. Heterozygote.

*Genotype.* A group or class sharing a specific genetic constitution, e.g., *XX* or *Xx*.
*Heterozygote.* Member of a genotype carrying both genes, dominant and recessive.
*Homozygote.* Member of a genotype carrying only one kind of gene, dominant or recessive.
*Mean Score.* The arithmetic mean of scores assigned to a number of events.
*Modal Score.* Score most frequently observed.
*Normalization Condition.* The requirement that all probabilities sum up to one.
*Probability Distribution.* Probability of an event or score as a function of the number *N* characterizing that event or score.
*Score.* Arbitrary number assigned to an event.
*Tree Diagrams.* Schematic representation of a sequence of events in which each branch represents a possible event.

## Problems

*1.* Draw the tree diagram for the sequential or simultaneous toss of two dice and draw the corresponding histogram.

*2.* Consider rolling three dice for a success defined as 5 on any die. What are the probabilities of 0, 1, 2, or 3 successes on one throw?

*3.* What is the probability of rolling two fives in a row (total spots on both dice) using two dice?

*4.* What is the probability that, in a family of five there will be 0, 1, 2, 3, 4, or 5 boys?

*5.* Show that Eq. (5-29) follows from Eq. (5-28).

*6.* Suppose a vaccine has a 0.9 probability of taking and three individuals are vaccinated. Calculate $f_{3,0.1}$, the predicted frequency for 0, 1, 2, and 3 vaccination failures. What is the most probable number of failures for 30 individuals?

*7.* Prove $\bar{n} = Np$ (difficult; see Young, 1962, Appendix B).

*8.* Prove $\sigma^2 = Npq$ (difficult; see Young, 1962, Appendix B).

*9.* Suppose a disease is known, by past experience, to be fatal in laboratory animals 80% of the time. Three animals are successfully infected with the disease, treated by a new method, and all three survive. Given only the data we have, what decision shall we make with regard to the significance of the three successful treatments and with what confidence may we make it, i.e., what is our probability of being right?

*10.* Suppose a new treatment is instituted for treatment of a disease with a 30% mortality rate and the first three of four diagnosed cases survive under the new treatment. Is this evidence for or against the new treatment, and what is the probability that the 3 of 4 survivals owe to chance alone?

*11.* In Problem 11, Chapter 4, we asked the probability of drawing two and only two green beads in three draws from a large (limitless) supply with p(G) = 0.20 and found that the probability of p(G,G,W) or (G,W,G or W,G,G) = 0.096. In Table 5-1, we have p(T,T,H or T,H,T or H,T,T) = 3/8 = 0.375. The problem is essentially the same but for the probability of a successful single draw or throw which was 0.20 in the first case and 0.50 in the second. Since 0.20/0.50 ≠ 0.096/0.375, the relationship between p of the first draw and p of two successful draws out of three is not linear. Write a computer program to determine p(2 of 3) vs p(first draw) over all possible values and plot the curve.

## Bibliography

W. S. Dorn, H. J. Greenberg, and S. K. Keller, *Mathematical Logic and Probability with BASIC Programming*, Prindle, Webber, and Schmidt, Boston, Mass. 1973.

H. D. Young, *Statistical Treatment of Experimental Data*, Wiley, New York, 1962.

# Chapter 6

# Using the Poisson Distribution

The Poisson distribution governs unlikely random events. An example of an unlikely random event is the radioactive disintegration of a uranium $^{238}_{92}U$ nucleus. The probability of such a disintegration is so small that an observer watching a single $^{238}_{92}U$ nucleus for the entire time the earth has existed would have only a 50-50 chance of seeing it disintegrate. The number of atoms in a milligram of $^{238}_{92}U$ is so huge, however, that numerous atoms decay in only a second of observation time. Just how numerous is, of course, a matter of chance. The Poisson distribution gives the probabilities associated with 1, 2, 3,. . . disintegrations.

The first part of this chapter shows that, although they have different mathematical forms, the Poisson distribution is a special case of the binomial distribution. The second part of the chapter gives applications of the Poisson distribution, while the third part of the chapter is an elementary introduction to the field of computer graphics—how to make the computer draw pictures and graphs.

If we observe radioactive disintegration per unit time or the number of bacterial colonies per unit area of an agar plate, the number of nuclei that can suffer disintegration is of the order of $10^{20}$ or more and the number of possible points on a plate is infinite. The probability of either a specific nucleus disintegrating or a colony appearing at a specific point is very small. We shall consider the behavior of the binomial distribution, $f_{N,p}$ $(n)$ where $p$ is very small and $N$ is very large or infinite. The mean of $n$, $\bar{n} = Np$, is finite and constant. Let us define it

$$Np \equiv \lambda \tag{6-1}$$

Since $p$ is small, $n << N$. Combination of Eqs. (4-43) and (4-45) shows that

$$\frac{N!}{(N-n)!} = N(N-1)(N-2)\ .\ .\ .\ (N-n+1) \tag{6-2}$$

For a very large $N$, subtracting 1, 2, 3 or any small number does not appreciably change the population

$$N \cong (N - 1) \cong (N - 2) \ldots \tag{6-3}$$

so that the product on the right of Eq. (6-2) is approximately

$$N \times N \times N \ldots \cong N^n \tag{6-4}$$

Now, the binomial frequency,

$$f_{N,p}(n) = \frac{N!}{(N - n)!n!} p^n q^{N - n} \tag{6-5}$$

can be simplified to give

$$f_{N,p}(n) = \frac{N^n}{n!} p^n (1 - p)^{N - n} = \frac{(Np)^n}{n!} (1 - p)^{N - n}$$
$$= \frac{(Np)^n}{n!} \frac{(1 - p)^N}{(1 - p)^n} \tag{6-6}$$

but $p$ is very small, hence for a relatively small $n$,

$$(1 - p)^n \cong 1^n = 1 \tag{6-7}$$

We cannot do the same thing with $(1 - p)^N$ because a number differing very slightly from 1 taken to a very large exponent is not equal to 1; that approximation holds only if the exponent is of reasonably small size. Now,

$$f_{N,p}(n) = \frac{(Np)^n}{n!} (1 - p)^N \tag{6-8}$$

but

$$\lambda = Np$$

hence

$$N = \lambda/p \tag{6-9}$$

and

$$f_\lambda(n) = \frac{\lambda^n}{n!} (1 - p)^{\lambda/p}$$
$$= \frac{\lambda^n}{n!} \left[(1 - p)^{1/p}\right]^\lambda \tag{6-10}$$

where we have substituted the notation $f_\lambda(n)$ indicating a Poisson distribution frequency for a specific Poisson coefficient, $\lambda$, which results from the binomial distribution $f_{N,p}(n)$, *under the conditions stated.*

It can be proven that

$$\lim_{p \to 0} \quad (1 - p)^{1/p} = 1/e \tag{6-11}$$

where the notation lim indicates that the *limit* of the quantity following, in this case, $(1 - p)^{1/p}$, approaches $1/e$ as $p$ approaches zero. The number $e$ is known as the base of the natural logarithm system, but we may regard it simply as a parameter, $e = 2.718$.

Now,

$$f_\lambda(n) = \frac{\lambda^n}{n!} (1/e)^\lambda = \frac{\lambda^n e^{-\lambda}}{n!} \tag{6-12}$$

We have gone to considerable algebraic trouble in applying these simplifying assumptions to obtain Eq. (6-12), but we have been repaid with a mathematical statement that is considerably easier to apply than the binomial distribution.

Summing all predicted frequencies, $f_\lambda\ (n)$ from zero to infinity,

$$\sum_0^\infty f_\lambda\ (n) = e^{-\lambda} \sum_0^\infty \frac{\lambda^n}{n!} \tag{6-13}$$

but it can be shown that

$$\sum_0^\infty \frac{\lambda^n}{n!} = e^\lambda$$

Consequently

$$\sum_0^\infty f_\lambda(n) = e^\lambda e^{-\lambda} = 1 \tag{6-14}$$

which is what we should expect.

When the sum of all values for a discontinuous probability function (or the integral for a continuous function) is one, the function is said to be normalized to one. Functions in general can be normalized to any number, but probability functions are always normalized to one.

## Application: Disintegration of Radioactive Nuclei

Radioactive disintegration of an individual atom of uranium, $^{238}_{92}U$, is not a very likely event. Uranium has a half-life of $4.51 \times 10^9$ years, which means that if we were to observe an individual atom of $^{238}_{92}U$ for four and a half billion years, we would have about a 50–50 chance of seeing it disintegrate.

Another way of interpreting the half-life is to recall from general chemistry (see also Chapter 10) that

$$t_{1/2} = 0.693/k \qquad (6\text{-}15)$$

where $k$ is the rate constant for the first-order radioactive decay

$$^{238}_{92}U \rightarrow {}^{234}_{90}Th + \alpha$$

and $\alpha$ represents an alpha particle that can be counted using a suitable counter. The rate law is

$$-\,dN/dt = kN \qquad (6\text{-}16)$$

where $N$ is the number of atoms present in a large collection, say a 0.0500 milligram sample. The negative differential term, $-dN/dt$, represents a decrease in $N$ because every $\alpha$ particle ejected by the sample means that one atom of uranium has been lost from the original population of $N$ atoms.

A sample of uranium weighing 0.0500 mg contains $5.00 \times 10^{-5}/238 = 2.10 \times 10^{-7}$ moles and, from general chemistry, we recall that one mole of atoms is $6.02 \times 10^{23}$ atoms, leading to

$$(2.10 \times 10^{-7})(6.02 \times 10^{23}) \cong 1.3 \times 10^{17}$$

atoms in the uranium sample we have chosen for observation.

The rate constant is, from Eq. 6-15,

$$0.693/4.51 \times 10^9 = 1.54 \times 10^{-10}\ \text{yr}^{-1}$$

We expect to observe

$$-\,dN/dt = kN$$
$$= 1.54 \times 10^{-10}(1.3 \times 10^{17}) \cong 2.0 \times 10^7 \text{ emissions yr}^{-1}$$

This seems like a very large number, but there are $10^{17} - (2.0 \times 10^7) \cong 10^{17}$ atoms that didn't disintegrate. The ratio of the number of events divided by the number of possibilities, $2.0 \times 10^7/1.3 \times 10^{17}$, gives the probability, $p = 1.54 \times 10^{-10}$, for any one event occurring in that year. We see that the probability of decay in one year is just the rate constant in units of $\text{yr}^{-1}$.

Suppose we imagine a series of α counting experiments, each of one second duration, and we wish to know the probability of observing 0, 1, 2, or 3 counts during that second. The computation is similar to the way that we computed the probability or predicted frequency for the binomial distribution.

The half-life, converted to seconds, yields the rate constant, $k = 4.87 \times 10^{-18} s^{-1}$. The Poisson distribution parameter, λ, is $Np$ or $1.3 \times 10^{17} \times 4.87 \times 10^{-18} = 0.63$

$$f_{0.63}(0) = \frac{\lambda^n}{n!} e^{-\lambda} = 0.53$$

$$f_{0.63}(1) = 0.63(0.53) = 0.34$$

$$f_{0.63}(2) = 0.11$$

$$f_{0.63}(3) = \underline{0.02}$$

$$1.00 = \text{sum of all } f$$

### Exercise 6-1

Obtain the rate constant in units of $s^{-1}$ for ${}^{14}_{6}C$ given that its half-life is 5720 years.

## Mean, Variance, and Standard Deviation for the Poisson Distribution

We have stated that the mean, $\bar{n}$, equals $Np$ for the binomial distribution. Consequently, by our definition of λ,

$$\bar{n} = Np = \lambda \tag{6-17}$$

for the Poisson distribution. Since

$$\sigma^2 = Npq \tag{6-18}$$

for the binomial distribution, and $q \cong 1$ for the Poisson distribution, we have

$$\sigma^2 = Np = \lambda \tag{6-19}$$

for the Poisson distribution which leads to the standard deviation,

$$\sigma = (\mathrm{Np})^{1/2} = \lambda^{1/2} \tag{6-20}$$

The parameter λ is really a nonnormalized probability and may have units, in contrast to the normalized probability, which is unitless. As we have developed the Poisson distribution thus far, it has been in terms of the number of discrete events occurring out of a number of discrete possible events, e.g., the number of atoms undergoing radioactive disintegration per unit time relative to the number of atoms present in a sample, each of which is capable of disintegrating. The units (atoms/atoms) cancel, making $p$ unitless.

Suppose, however, that we were doing a survey of the migration patterns of fish in the gulf stream and we had determined an average population of three fish of a certain kind over a large area to be surveyed. This average is not $p$, but $\lambda$ in the Poisson distribution. The parameters $N$ and $p$ are not defined in this situation because there is no way of assigning the number of "potential fish" to a cubic mile of water; a fish is either there or it is not.

When the Poisson distribution is computed, $f(0)$, $f(1)$, . . . are expected frequencies or normalized probabilities that a cubic mile of water will be found with no fish, one fish, . . . etc.

## Exercise 6-2

Suppose that the average rate of attack upon blue spruce by a certain bark beetle is three trees per square mile. A survey is done to determine whether a bark beetle infestation exists in several national forests by randomly surveying square mile segments, and counting the number of trees in each attacked by the bark beetle. The decision whether or not to cut a stand of blue spruce is made by comparing its rate of infestation to the normal rate over all forests. If a stand shows significantly higher infestation than normal, it is cut to prevent spreading of the infestation. What are the probabilities of finding zero, one, two, . . . attacked trees in a square mile or normal stand?

Solution 6-2. From the Poisson distribution with $\lambda = 3$

Table 6-1

| $n$ | $f_\lambda(n)$ | $n$ | $f_\lambda(n)$ |
|---|---|---|---|
| 0 | 0.050 | 5 | 0.101 |
| 1 | 0.149 | 6 | 0.050 |
| 2 | 0.224 | 7 | 0.021 |
| 3 | 0.224 | 8 | 0.008 |
| 4 | 0.168 | 9 | 0.002 |

## Computer Graphics

One of the most fascinating applications of computers is that of computer graphics, printing out sets of symbols according to some pattern or design. There is no limit to the designs that can be printed out by a computer except in the ingenuity of the individual programmer. Computers can be programmed to print out very literal pictures, resembling, from a distance, black and white photographs. Pictures of Einstein and young ladies surprised, presumably while preparing for a trip to the beach, are popular, but they require long programs. We shall restrict ourselves to elementary problems, one in this chapter for filling in areas and one for drawing lines in the next.

### Programs 6-1 and 6-1A

We wish to write a program that prints out asterisks (or other symbols) for every value of *n* less than some input or calculated value. In this way, computer print-out can be produced that has areas filled in or left blank according to the type of generating function controlling the value of *n*. This is a useful visual technique for representing histograms and model or ideal frequency distributions. The program will be applied to the Poisson distribution in this chapter and the normal frequency distribution in the next.

```
10   FOR I = 1 TO 10
20   PRINT"*";
30   NEXT I
40   END

READY
RUNNH


**********

TIME:   0.08 SECS.
```

```
10   FOR O = 1 TO 10
20   PRINT
30   FOR I = 1 TO 10
40   PRINT"*";
50   NEXT I
60   NEXT O
70   END

READY
RUNNH


**********
**********
**********
**********
**********
**********
**********
**********
**********
**********

TIME:   0.23 SECS.
```

**Commentary on Programs 6-1 and 6-1A.** Program 6-1 prints a line of asterisks (or other symbols) using a FOR–NEXT loop containing an alphabetic * followed by a semicolon. Without the semicolon that prevents the printer from proceding to a new line on each iteration of the loop, this program would give a vertical line of symbols rather than a horizontal line. Subsequent programs will make it evident why a horizontal line is desirable for our purposes.

Modification 6-1A shows the entire program 6-1 included within an outer loop (the O loop) that iterates the inner or I loop ten times, leading to ten vertical lines of ten asterisks each. We now have a method of producing a 10 × 10 area populated by 100 asterisks. The 20 PRINT statement is said to "bump" carriage control off the line it is on so that a new line is started on each iteration of the outer loop.

## Program 6-1B

```
5    N=0
10   FOR J= 1 TO 10
15   N=N+1
20   PRINT
30   FOR I = 1 TO N
40   PRINT"*";
50   NEXT I
60   NEXT J
70   END

READY
RUNNH


*
**
***
****
*****
******
*******
********
*********
**********

TIME:  0.17 SECS.
```

**Commentary on Program 6-1B** In modification 6-1B, N is initialized at zero before entering the outer loop and incremented by one on each iteration of the outer loop. (The running index of the outer loop has been changed from O to J to emphasize that it is merely a counter and carries no special significance.) The inner loop has been changed to run from 1 to N, hence on the first pass through the inner loop, with N = 1, there is only one iteration and only one asterisk is printed. On the second iteration of the outer loop, N = 2 and the inner loop prints **. This continues for ten iterations of the outer loop producing a triangular area which, when rotated

90° to the left, is the area bounded by a horizontal, a vertical, and the linear function $y = mx$. The slope of the area mapped out is not equal to one because the scale of the axes are not the same; the space between lines is larger than the space between characters.

## Program 6-1C

```
5    N=0
6    X=0
10   FOR J = 1 TO 35
12   X=X+0.2
15   N=35+30*SIN(X)
20   PRINT
30   FOR I = 1 TO N
40   PRINT"*";
50   NEXT I
60   NEXT J
70   END

READY
RUNNH
```

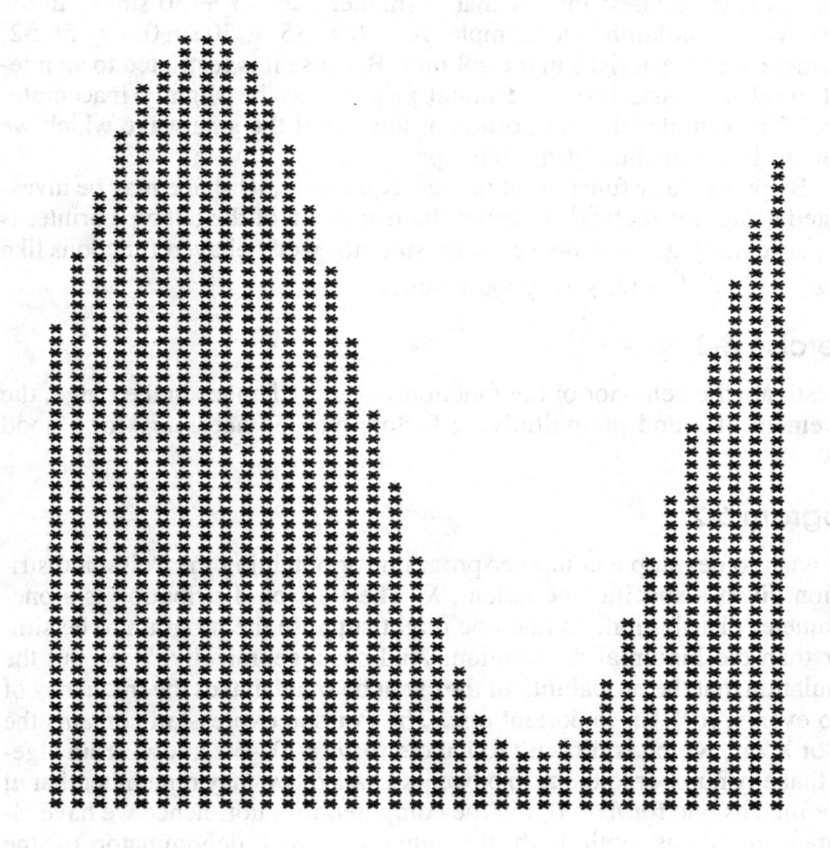

```
TIME:  1.41 SECS.
```

**Commentary on Program 6-1C.** Modification 6-1C shows one way of filling in the area under a nonlinear function, in this case a sine function. The values of N and X are initialized at zero, and the J loop is set for 35 iterations, permitting us to investigate a greater range of the function than in the previous modification. Statement 12 increments X by 0.2 and statement 15 sets N equal to 35 + 30 sin X where we know that sin X will not be more than 1 nor less than −1. Thus, the premultiplier 30 gives us an amplitude of −30 to +30. The additive constant 35 is used to keep the entire function on the page; we know that N will never be less than 5 nor more than 65. If uneven line spacing appears near the peak of the output graph, reduce this constant to 32. After the value of N has been set equal to a sine function, steps 20 through 70 are exactly as they were in the previous modification. Asterisks are printed up to but not beyond the value of N. Because only an integral number of asterisks can be printed, the number of asterisks is the nearest integer that is smaller than 35 + 30 sin X. In the fourth vertical column, for example, X = 0.8, 35 + 30 sin 0.8 = 56.52, and there are 56 asterisks in the column. Because it is restricted to an integral number of asterisks, a terminal printer may be slightly inaccurate. Note, for example, the flat portion at the top of the sine wave which we know to be an artifact of the teletype.

By putting any function of interest equal to N, its shape may be investigated using this method. Because the resolution of the teletype printer is not particularly good, however, only smooth, well behaved functions like $x^2$, $e^x$, $e^{-x}$, $e^{-x^2}$, etc. give good results.

## Exercise 6-3

Investigate the behavior of the function $y = x^2$. Changes in the range, the increment of $x$ and premultiplying factors may be necessary for a good plot.

## Program 6-2

We wish to develop a computer program for obtaining the Poisson distribution given a specific coefficient, $\lambda$. The Poisson distribution is a one-parameter distribution, so that one might suppose the program to be simpler than the binomial distribution. And so, algebraically, it is, but the stipulation that the probability of the event be small makes the frequency of zero events, $f(0)$, an important one. The Poisson expression contains the factor $\lambda^n/n!$, which is difficult to handle for $n = 0$. We know, from algebra that any number to the zero power is 1. Also, we may remember that $n!$ is defined as one for $n = 0$, but the computer does not, hence we have algebraic problems with both the numerator and denominator of the premultiplier in the Poisson distribution for $n = 0$. The difficulty is circumvented by constructing the program with two compute modules, one for $f(0) = e^{-\lambda}$ and the second for the remaining frequencies in the distribution

$$f_\lambda(n) = (\lambda^n/n!)e^{-\lambda}$$

Program 6-2 is illustrated taking $\lambda = 2.0$.

```
10  L=2
20  F=EXP(-L)
30  PRINT F
40  F1=1
50  N=1
60  F1=F1*N
70  F=L**N*EXP(-L)/F1
80  PRINT F
90  N=N+1
100 IF N<10, THEN 60
110 END

READY
RUNNH

 0.135335
 0.270671
 0.270671
 0.180447
 9.02235E-2
 3.60894E-2
 1.20298E-2
 3.43709E-3
 8.59272E-4
 1.90949E-4


TIME:  0.22 SECS.
```

## Exercise 6-4

Write the commentary for Program 6-2.

## Program 6-3

We wish to write a program combining the computer graphics technique with Program 6-2 as the generating function so as to display the Poisson distribution in graphical form. The program as written generates the Poisson distribution for integral values of $\lambda$ where $\lambda$ is represented by L in the program. Selected output graphs are shown for $\lambda = 2, 5$, and 10. As $\lambda$ becomes larger, the Poisson distribution broadens out and approaches a *normal distribution*, the subject of the next chapter. The printout has been rotated 90° to the left.

```
1    L=2
2    N=0
3    F=250*EXP(-L)
4    FOR I=1 TO F
5    PRINT"*";
6    NEXT I
7    REM  ***END OF MODULE FOR N=0***
8    F1=1
10   FOR J=1 TO 20
13   N=N+1
14   F1=F1*N
15   F=(L**N*EXP(-L))/F1
16   F=250*F
20   PRINT
30   FOR I=1 TO F
40   PRINT"*";
50   NEXT I
60   NEXT J
70   END

READY
RUNNH
```

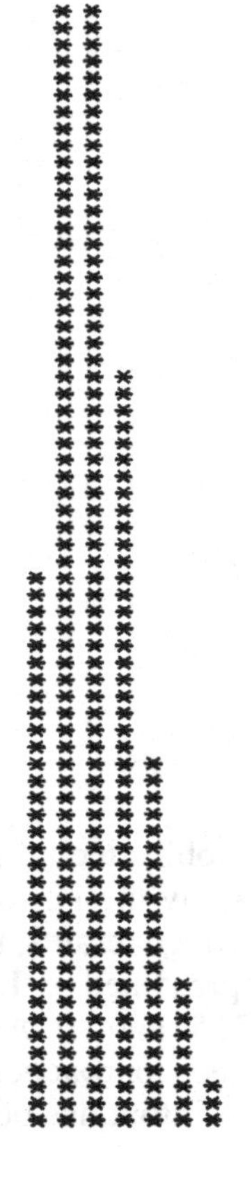

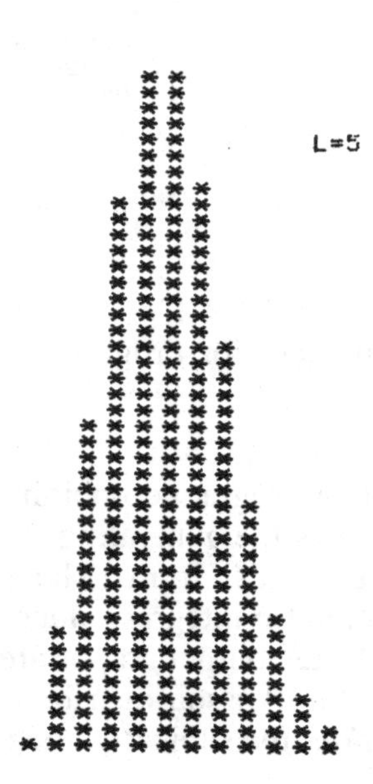

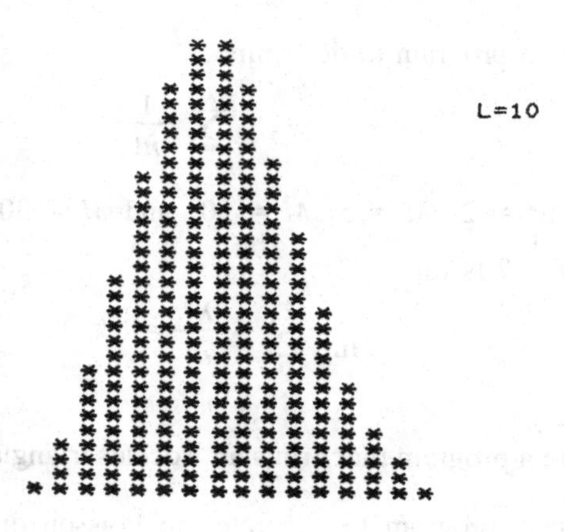


### Exercise 6-5

Write the commentary for Program 6-3. Note the use of a REM statement for clarification. For CRT screens, it may be necessary to scale down the plot by changing statement 16 to F = 200*F.

## Glossary

*Poisson Distribution.* Statistical distribution governing rare events.
*Normalization.* Setting $\Sigma f(n)$ equal to some arbitrary number, frequently one.
*Half-Life.* Time necessary for one half of a set of statistically governed events to occur, e.g., the time needed for one half of the nuclei of a radioactive sample to decompose.

## Problems

*1.* Prove that

$$\lim_{p \to 0}(1 - p)^{1/p} = 1/e$$

*2.* Prove that

$$\Sigma \frac{\lambda^n}{n!} = e^{\lambda}$$

*3.* If the probability that an individual will die from a certain disease is 0.005, what is the probability that out of 1000 people having this disease, exactly 4 will die?

*4*. Write a program to determine

$$\sum_{n=1}^{M} \frac{1}{n!}$$

for $M = 1$, $M = 2$, $M = 5$, $M = 20$, and $M = 30$. What is the limit of the sum $\Sigma\frac{1}{n!}$? Is lim

$$\lim \sum_{0}^{\infty} \frac{\lambda^n}{n!} = e^{\lambda}$$

for $\lambda = 2$?

*5*. Write a program that prints an isoceles triangle filled with question marks.

*6*. Write a program to generate the Poisson distribution for $\lambda = 1$ through 10. Draw all ten curves on the same graph paper using the same coordinates. Modify the program to generate the distribution for $\lambda = 0.5$ through 9.5. Draw all ten curves as before.

*7*. For observation of radioactive disintegrations in a sample of an element containing $10^{13}$ atoms, $f(n)$ has a maximum at about $n = 2$ counts per second over many 1-s counting periods. What is the approximate half life in years for the element?

*8*. Indium has a radioactive isotope $^{116}$In that decays with a half life of 54.2 min. If a 65 mg sample of In has a count rate of 50,000 counts per minute, how many radioactive atoms are present? What fraction of the sample is radioactive? What is the mean number of disintegrations expected in a series of observations, each of one second duration? What is the anticipated standard deviation about the mean?

*9*. Prove that the mean lifetime $\tau$ of a collection of radioactive nuclei is $\tau = 1/k = t_{1/2}/0.693$.

## Bibliography

W. C. Schefler, *Statistics for the Biological Sciences*, Addison-Wesley, Reading, Mass., 1969.

H. D. Young, *Statistical Treatment of Experimental Data*, McGraw-Hill, New York, 1962.

# Chapter 7

# Using the Normal Distribution

The normal distribution is the best-known of the statistical distributions. It results in the familiar "bell-shaped" or Gaussian curve. In this chapter, we shall develop a relationship between the Gaussian distribution and the mean and standard deviation treated earlier. Once we know the population mean and standard deviation of a randomly distributed continuous variable, we can predict the probability that future measurements will fall within any arbitrarily defined interval of the range of all possible measurements. We can also make statistical judgments whether any individual measurement or observation belongs to the set with a given mean and standard deviation or whether it belongs to some other set. These judgments rely on integration of the Gaussian function. In this book, of course, we do our integration by computer, a technique that is discussed at some length near the end of the chapter.

If the frequency of occurrence of a randomly distributed continuous variable $x$ is observed very many, strictly speaking, infinitely many times, the curve shown in Fig. 7-1 results. Figure 7-1 is the limiting case of binomial histograms, as shown in the progression from (a) to (c) in Fig. 7-2.

The Gaussian distribution curve may be fit by the empirical equation

$$f(x) = A \{\exp - h^2(x - m)^2\} \tag{7-1}$$

where $f(x)$ is a frequency function describing observed outcomes of events $x$ falling in a narrow interval of the entire range of $x$, $e$ is the natural base 2.718, and $A$, $h$, and $m$ are empirical parameters. The subject of empirical curve fitting is to be taken up in detail in later chapters. Suffice it to say here that, just as it is always possible to draw the curve representing an equation, it is also always possible to obtain an equation that represents a given curve, for example, a graph of experimental data. The equation obtained is said to be an *analytical expression* of the experimental curve that is a *graphical representation* of the data. In the process of obtaining the analytical expression of a curve, several of numbers appear that are as-

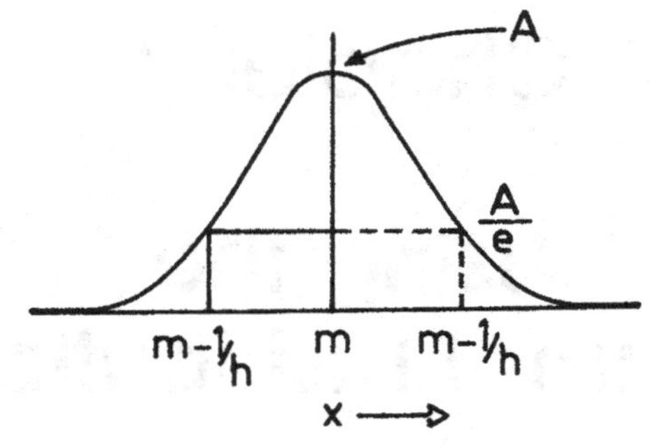


Fig. 7-1. The normal or Gaussian frequency distribution.

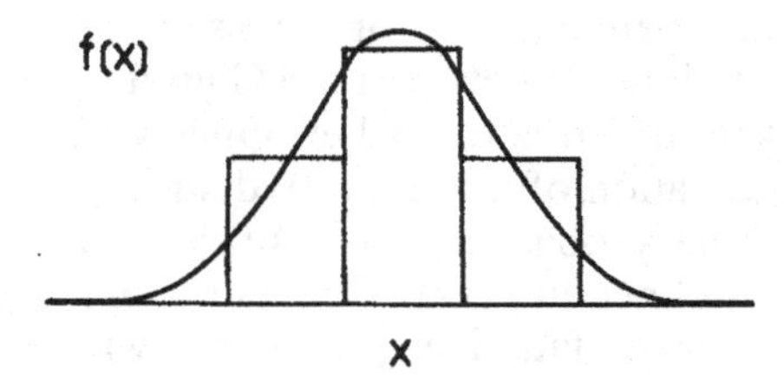


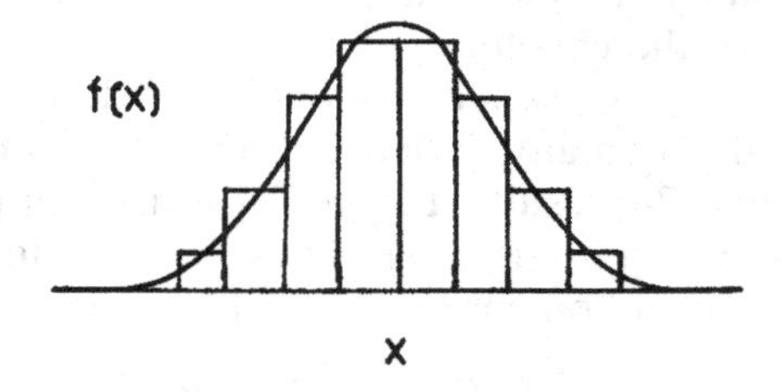


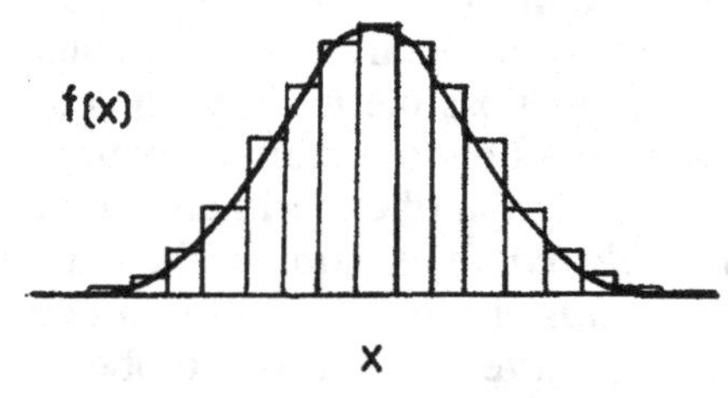


Fig. 7-2. Binomial histograms approaching the Gaussian curve.

signed arbitrary values so as to make the equation fit the curve. These are called *empirical parameters* with the examples $A$, $h$, and $m$ in the analytical expression of the Gaussian curve, Eq. (7-1). Further analysis of the parameters obtained by simple empirical curve-fitting often yields a theoretical interpretation of them.

In the case of Eq. (7-1), which gives the distribution of a random variable about a central point, $m$, it should be evident that $m$ is the arithmetic mean and $1/h$, which is large for a large experimental scatter and small for a small experimental scatter, is related to the standard deviation (though they are not equal).

The expected frequency of observing a continuously distributed variable, $x$, to fall in the infinitesimal interval $x + dx$ is $f(x)dx$. When dealing with a continuous variable distributed over infinitesimal intervals $x + dx$, we may represent the sum of all frequencies over some *finite* interval, $a$ to $b$, as the integral

$$f(a,b) = \int_a^b f(x)dx \tag{7-2}$$

where $f(a,b)$ is the expected frequency of observing events in the finite interval $a$ to $b$. The integral sign $\int$ now replaces the summation sign encountered in the section on cumulative frequency distributions. In fact, Eq. (7-2) is still a summation, except that because of the infinitesimal size of the intervals, $dx$, it must be an infinite sum over any finite interval, $a$ to $b$. A simple way to obtain the value of an integral of a function over a finite interval is to measure the area under the curve over that interval. This is done in Fig. 7-3 for the expected frequency of observing experimental values of the variable $x$ that fall within the interval $a$ to $b$.

Measurement of the area under the curve may be done by quite crude means. The curve may be drawn on graph paper and its area may be estimated by counting squares under the curve within the interval, guessing at the sum of squares that have been cut in two by the curve. Alternatively, the curve can be drawn on graph paper and the portion of interest within

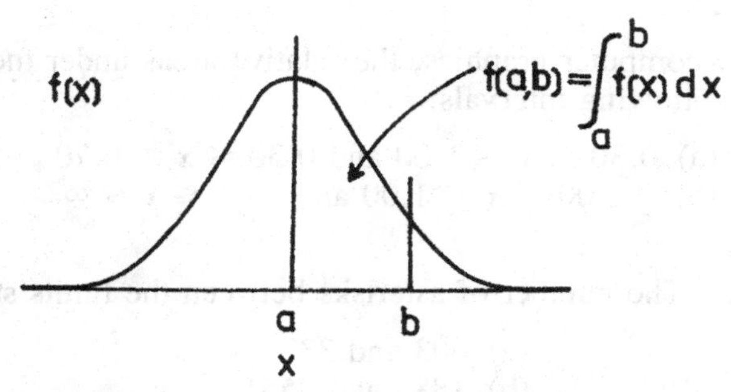


Fig. 7-3. Graphical integration by taking the area under a Gaussian curve.

the interval may be cut with a scissors and weighed. The ratio of the cut portion to that of a rectangle of known area gives the approximate area under the curve within the interval.

The area of any interval of a Gaussian curve is not the most useful information we can have because it gives a number that is proportional to, but not equal to, $f(a,b)$. Thus, by measuring the areas of two intervals we obtain the ratio of $f(a,b)$ to $f(c,d)$, but not the absolute values of either expected frequency. To obtain absolute values, we must perform some kind of normalization. One approximate graphical way would be to cut out the entire Gaussian curve and assign to this weight an arbitrary value of one. Then, by cutting and weighing the portion over the interval chosen and taking that weight relative to the first, we have the absolute value of $f(a,b)$.

### Exercise 7-1

Bearing in mind that the normal distribution is the limiting case of the binomial distribution, refer back to Chapter 5 and guess what $A$, $h$, and $m$ are in terms of $n$, $p$, and $q$. The answer to this question will emerge as the chapter progresses.

## Relative Areas by Computer Graphics

Rough comparison of areas under a Gaussian curve may be made by using the generating function N = 100*EXP(−Z*Z/2.) inside the FOR–NEXT loop of the area plotting program given in the last chapter (modification 6-1C). If Z is initialized at −3.0 outside the DO loop and incremented by 0.1 inside it, the Gaussian is printed as shown in Fig. 7-4. To approximate the area ratio over two intervals, one need merely count asterisks from the lower to the upper limits of the intervals and take the ratio of their number. The method is illustrated in the following example.

### Exercise 7-2

Determine, by computer graphics, the relative areas under the Gaussian curve for the following intervals.

(a) $0.30 \leq x \leq 1.00$ and $0.30 \leq x \leq 0.70$
(b) $-1.00 \leq x \leq 1.00$ and $-\infty \leq x \leq \infty$

Solution 7-2. The number of asterisks between the limits stated are

(a) 603 and 337
(b) 1883 and 2551

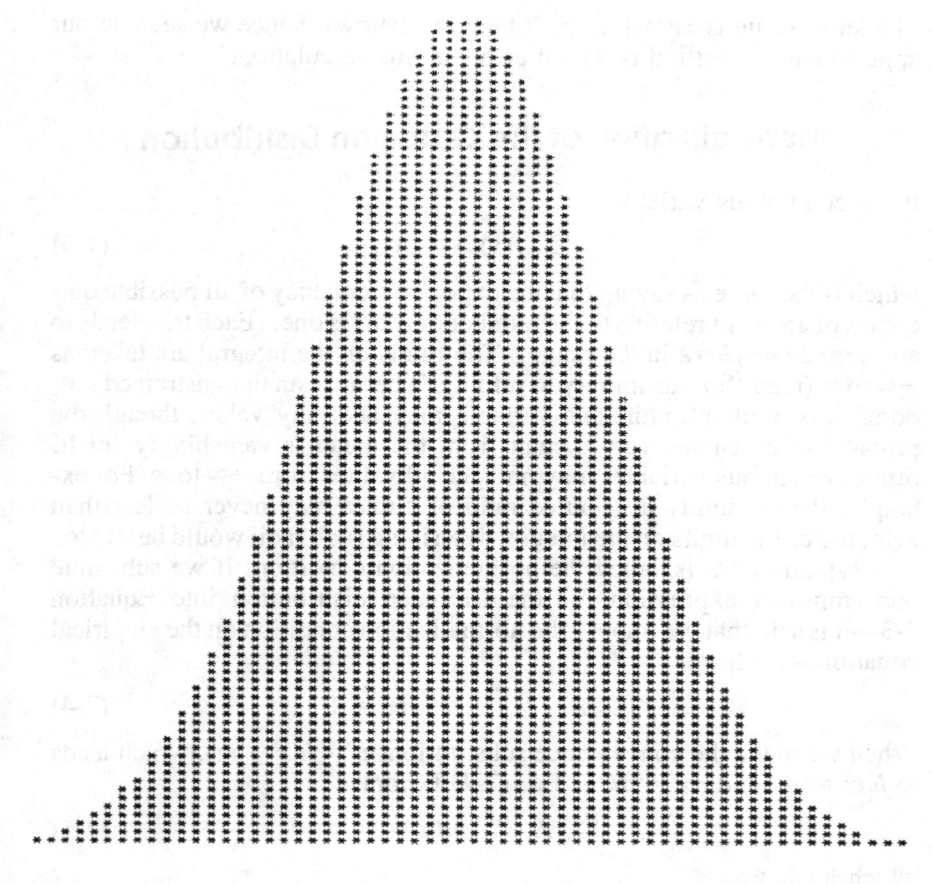

Fig. 7-4. Output for an area plot of the Gaussian function.

which lead to the approximate ratios of areas

(a) 1.79:1
(b) 0.738:1

Counting asterisks is a crude form of integration and the ratios obtained approximate the ratios of the definite integrals

$$\text{(a)} \int_{0.30}^{1.00} e^{-z^2/2}\, dz \Big/ \int_{0.30}^{0.70} e^{-z^2/2}\, dz \cong 1.79$$

and

$$\text{(b)} \int_{-1.00}^{1.00} e^{-z^2/2}\, dz \Big/ \int_{-\infty}^{\infty} e^{-z^2/2}\, dz \cong 0.738$$

The latter value is known to be 0.683 (see below), hence we see that our approximation method is 8% in error on this calculation.

## Normalization of the Gaussian Distribution

For a continuous variable,

$$\int_{-\infty}^{\infty} f(x)dx = 1 \qquad (7\text{-}3)$$

which is the same as saying that the expected frequency of all possible outcomes of an event relative to the number of trials is one. (Each trial leads to an event *somewhere* in the range.) The limits on the integral are taken as $-\infty$ to $\infty$ (read "minus infinity to infinity") because an unconstrained randomly distributed continuous variable may take any value, though the probability of values very distant from the mean is vanishingly small. Some continuous variables may not take all values from $-\infty$ to $\infty$. For example, the randomly distributed radii of circles may never be less than zero; hence the limits on the integral describing such radii would be 0 to $\infty$.

Equation 7-3 is called the normalization condition. If we substitute our empirical expression for the Gaussian distribution into Equation 7-3—it is said that we impose the normalization condition on the empirical equation—we have

$$\int_{-\infty}^{\infty} Ae^{-h^2(x-m)^2} dx = 1 \qquad (7\text{-}4)$$

When we make the convenience substitutions $u = h(x - m)$, which leads to $hdx = du$ and $dx = du/h$, Eq. (7-4) becomes

$$\int_{-\infty}^{\infty} Ae^{-u^2} du/h = 1 \qquad (7\text{-}5)$$

which leads to

$$A\int_{-\infty}^{\infty} e^{-u^2} du = h \qquad (7\text{-}6)$$

Looking into a table of integrals, we find that the integral in Eq. (7-6) is known and is

$$\int_{-\infty}^{\infty} e^{-u^2} du = \pi^{1/2} \qquad (7\text{-}7)$$

This leads to

$$A(\pi)^{1/2} = h \qquad (7\text{-}8)$$

and we have evaluated $A$, called the normalization constant. It is $A = h/\pi^{1/2}$ which leads to

$$f(x) = (h/\pi^{1/2})\, e^{-h^2(x-m)^2} \qquad (7\text{-}9)$$

If Eq. (7-9) is plotted, the area under the curve over any interval $a$ to $b$ is the absolute predicted frequency of events falling in that interval, relative to all possible events because Eq. (7-9) has been normalized to one. Normalization, therefore, is seen as a mathematical technique for setting the entire area under the curve equal to one, whereupon any finite subsection of that area must be greater than zero and less than one, the condition that we demand of all probabilities.

### Exercise 7-3

Another view of normalized and nonnormalized data is given by returning to a simple integral data set in the following problem. Calculate the mean of (a) the ungrouped integer data 4, 3, 4, 5, 3, 4, 4, and (b) the grouped (nonnormalized) data shown in Table 7-1 and (c) the grouped normalized data in Table 7-2.

Solution 7-3. (a) The mean of the ungrouped data is

$$\bar{x} = \sum x_i / N = 27/7 = 3.86$$

(b) The mean of the grouped data in Table 7-1 is

$$\bar{x} = \sum x f(x) / N = 27/7 = 3.86$$

(c) The mean of the grouped normalized data in Table 7-2 is

$$\begin{aligned} \bar{x} = \sum x f(x) &= 0.286(3) + 0.571(4) + 0.143(5) \\ &= 0.853 + 2.284 + 0.714 = 3.86 \end{aligned}$$

Table 7-1
Grouped Data Set
Given Above

| $n$ | $f(n)$ |
|---|---|
| 3 | 2 |
| 4 | 4 |
| 5 | 1 |

Table 7-2
Grouped, Normalized
Integral Data Given
in Table 7-1

| $n$ | $f(n)/N$ |
|---|---|
| 3 | 2/7 = 0.286 |
| 4 | 4/7 = 0.571 |
| 5 | 1/7 = 0.143 |
| | 1.000 = sum |

## The Mean, Variance, and Standard Deviation of a Gaussian Distribution

By analogy to a previous equation for the mean of a discontinuous distribution (Eq. 5-14 and Exercise 7-3),

$$\bar{n} = \sum nf(n) \tag{7-10}$$

we may write an expression for the mean of a continuous, Gaussian distribution

$$\bar{x} = \int_{-\infty}^{\infty} xf(x)dx = h\big/\pi^{1/2} \int_{-\infty}^{\infty} xe^{-h^2(x\,-\,m)^2}\,dx \tag{7-11}$$

where the constants $h/\pi^{1/2}$ have been taken outside the integral sign in precisely the way that one factors a constant from a sum. By a procedure that follows the same steps as the normalization procedure, but results in a slightly more cumbersome algebraic development (which we shall not give here), the integral can be reduced to

$$\bar{x} = m\big/\pi^{1/2} \int_{-\infty}^{\infty} e^{-u^2}\,du \tag{7-12}$$

We already know that the value of the integral is $\pi^{1/2}$; hence we have

$$\bar{x} = (m\big/\pi^{1/2})\,\pi^{1/2} = m \tag{7-13}$$

Eq. (7-13) is an explicit statement of a conclusion that we already supposed to be true on qualitative grounds; the peak of the Gaussian curve, given by the empirical parameter, $m$, is, in fact, the mean.

Our expression for the variance of the distribution is

$$\sigma^2 = \int_{-\infty}^{\infty} (x - m)^2 f(x)dx \tag{7-14}$$

which, after replacing $f(x)$ by the Gaussian and making convenience substitutions, leads to

$$\sigma^2 = 1/h^2\pi^{1/2} \int_{-\infty}^{\infty} u^2 e^{-u^2}\,du \tag{7-15}$$

Again, the value of the integral is known and we may look it up in tables of standard integrals to find

$$\int_{-\infty}^{\infty} u^2 e^{-u^2}\,du = \pi^{1/2}\big/2 \tag{7-16}$$

whence

$$\sigma^2 = 1\big/2h^2 \tag{7-17}$$

or

$$\sigma = 1\big/2^{1/2}h \tag{7-18}$$

Notice that, as we anticipated from our qualitative investigation of the curve, $\sigma$ is related to $1/h$. Both $\sigma$ and $h$ are measures of precision, $\sigma$ being small for narrowly grouped, precise data and $h$ being large.

Since $h = 1/2^{1/2}\,\sigma$

$$h^2 = 1/2\sigma^2 \qquad (7\text{-}19)$$

We have now determined the theoretical values of what had previously been mere empirical constants, $A$, $h$, and $m$. Substituting each into the Gaussian distribution,

$$f(x) = 1/(2\pi)^{1/2}\,\sigma\, e^{-(x\,-\,\bar{x})^2/2\sigma^2} \qquad (7\text{-}20)$$

We have gone through a great deal of mathematical manipulation to obtain an equation in its most refined form, but the labor has been worth it for Eq. (7-20) conveys much more information than the empirical form. The empirical constants in Eq. (7-1) must be calculated by trial and error, a tedious procedure if not done by computer, and one that gives no intuitive understanding of their meaning. By contrast, $\bar{x}$ and $\sigma$ are fundamental concepts in statistics, are readily calculable, and give the reader a good "feel" for how Eq. (7-20) is related to the Gaussian error curve in Fig. 7-1.

Once again we invoke the idea that the frequency of many past events is the probability or expected frequency of future events. Thus the Gaussian function, which has been developed as an empirical function describing many past observations, will be regarded as a *probability function* governing future events. The further development of this chapter will rest heavily on the idea that the probability of an event occurring within the interval [$a$, $b$] is given by the area under the Gaussian from $a$ to $b$. This is, in effect, the assumption that the event follows a normal distribution.

## The Standard Normal Deviate, *z*

Figure 7-5 shows four normal distribution curves plotted on the same axis; A and B have the same standard deviations but different means and C and D have the same means but different standard deviations.

A convenience substitution,

$$z = (x - \bar{x})/\sigma \qquad (7\text{-}21)$$

converts these fundamentally different Gaussian distributions to the same form, leading to

$$\int_{z_1}^{z_2} p(z) = \int_{z_1}^{z_2} 1/(2\pi)^{1/2}\, e^{-z^2/2}\, dz \qquad (7\text{-}22)$$

where the notation $p(z)$ indicates a probability.

We are now able to draw conclusions using one simple technique for any and all normal curves. The mode of attack is (1) convert the actual curve to the standard normal curve in terms of the standard normal deviate, $z$; (2) draw conclusions from the standard normal curve; and (3) translate

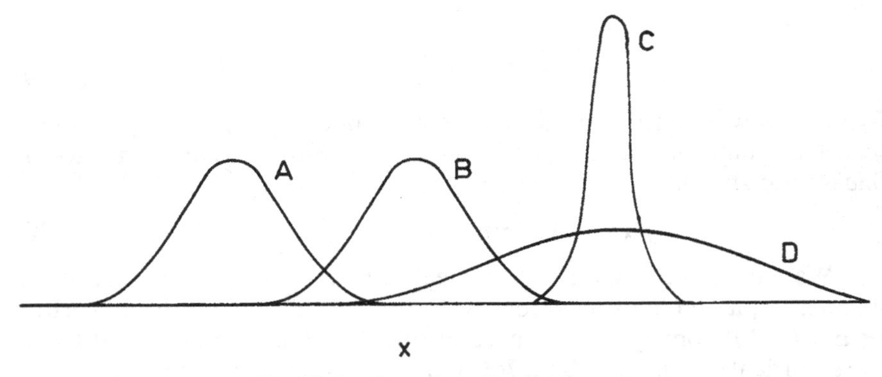


Fig. 7-5. Gaussian distributions.

the results in terms of the standard normal deviate $z$ back into the terms of the original problem.

## Exercise 7-4

Show that Eq. (7-21) results in (7-22) when substituted into (7-20), assuming that we can substitute $p(z)$ for $f(z)$ as discussed in the last section.

## Exercise 7-5

Bilirubin is a bile pigment present in the blood of healthy individuals, but elevated in the blood of those suffering from jaundice, hepatitis, and pernicious anemia. Suppose that analysis is performed on the blood serum samples of many healthy individuals and that the mean value is 0.90 mg/100 mL, and the standard deviation is 0.20 mg/100 mL. Assuming a normal distribution, what is the probability that the next analysis will yield a result outside the interval 0.90 ± 0.20 mg/100 mL *for a healthy individual*?

Solution 7-5. The limits of the range we are concerned with, are 0.70 and 1.10 mg/100 mL.

(1) Converting to the standard normal deviate (frequently called the "$z$ score")

$$z = \frac{0.70 - 0.90}{0.20} = -1$$

and

$$z = \frac{1.10 - 0.90}{0.20} = 1$$

(2) The area under a standard normal curve between the limits $-1$ and $+1$ is 0.6826. (The reader should accept this on an *ex cathedra* basis for the moment—we will investigate the areas under the standard normal curve in detail in a later section.) The total area under the curve is normalized to 1.0. The probability that any normally distributed event will be within the limits $z = \pm 1$ is 0.6826, or about 68.3%. The probability that $z$ will lie outside these limits *on a chance basis only* is $100 - 68.3 = 31.7\%$.

(3) Going back to the units of the original question, from the general statements just made about $z$, which apply to any normally distributed event, the probability that the next bilirubin analysis of a healthy individual will lie outside the range 0.70–1.10 mg/100 mL is 31.7%.

## Program 7-1

Computer graphics for curve-drawing are introduced by a single BASIC program, which is applied to the function $e^{-z^2/2}$. The parameters were chosen arbitrarily to make the curve fill a page of computer output. For display on CRT screens, the number of points can be reduced from 61 to 31, or 16 by reducing the upper limit in statement 20. Each reduction by half should be accompanied by an increase in incremental steps in statement 50, e.g., to 0.2 and 0.4 for each of the changes suggested above. Reducing the premultiplier 70 in statement 30 reduces the height of the output curve. Many other sealing factors are possible. Experiment with them.

```
10   Z=-3.
20   FOR I=1 TO 61
30   K=INT(70*EXP(-Z*Z/2.))
40   PRINT TAB(K)"*"
50   Z=Z+.1
60   NEXT I
70   END
```

**Commentary on Program 7-1.** Statement 10 sets the lower limit of the interval over which the function is to be graphed. The probability of events occurring three standard deviations below the mean is very small; hence, for graphing purposes, it may be considered to be zero. Statement 20 enters a loop that iterates its contents 61 times to obtain $3.0/0.1 = 30$ points for values of $z$ smaller than zero, 30 points for $z > 0$, plus one point at $z = 0$. Within the loop, K is set equal to the integer value of $70\ e^{-z^2/2}$ where the premultiplier 70 is arbitrarily selected to fill a page of teletype printout. The system subroutine INT truncates the value of any function within its parentheses to the integer just smaller than the true value of the function. Statement 40 operates as the TAB key on a typewriter does. TAB(K) skips K spaces from the left margin before printing whatever follows, in this case an asterisk. The function of the TAB state-

```
RUNNH

*
 *
 *
 *
  *
   *
   *
    *
      *
       *
         *
            *
              *
                *
                    *
                       *
                           *
                               *
                                   *
                                       *
                                           *
                                               *
                                                   *
                                                       *
                                                           *
                                                              *
                                                                 *
                                                                   *
                                                                     *
                                                                      *
                                                                      *
                                                                      *
                                                                     *
                                                                   *
                                                                 *
                                                              *
                                                           *
                                                       *
                                                   *
                                               *
                                           *
                                       *
                                   *
                               *
                           *
                       *
                    *
                 *
              *
            *
          *
        *
       *
     *
    *
    *
   *
  *
  *
  *
 *
```

ment makes it clear why K was specified as an integer variable; the teletype can only skip an integral number of spaces. After incrementing $z$ by 0.1, the loop is reiterated resulting in another asterisk at the integer below 70 $e^{-z^2/2}$. After the loop is complete, the entire 61 points have been printed as shown in the output. There are minor irregularities in the printed curve, most noticeable at the peak and both tails owing to the integer truncation of the function just discussed.

## Computer Integration

We saw that the area under the Gaussian curve was essential to the solution of Exercise 7-5 and we saw that a rough estimate of that area can be obtained by counting asterisks in the area plot, Fig. 7-4. Now let us refine the method so as to obtain the area between any two values of $z$.

Most integrals do not have closed solutions, but their functions can be graphed. Suppose we have the graph of an arbitrary function, $f(x)$, over an interval, $a$ to $b$. Suppose further that the function does not change much over the interval. We can estimate the area under the curve by determining the value of the function at some point, say $a$, and multiplying by the width of the interval, $ab$.

Clearly, the estimate obtained by this method is a poor one, and it is worse for functions that vary more steeply than the one we selected. However, like computer methods we have already seen and used, this unpromising method of estimation can be made profitable by putting it into an iterative routine and taking advantage of the computer's ability to perform repetitive computations at incredible speed.

The way we bring the computer into play in this problem is, not surprisingly, to break up area estimation into many small steps. Looking back to the function just considered, we can get a better approximation of the area under the curve by computing the area of two rectangles, one from $a$ to $(b - a)/2$ and the other from $(b - a)/2$ to $b$.

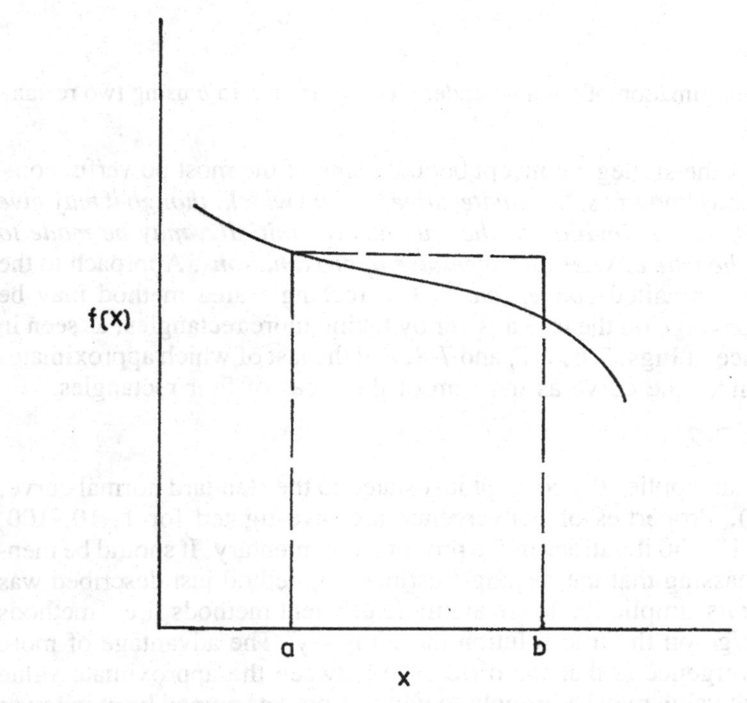


Fig. 7-6. Estimation of the area under $f(x)$ vs $x$ from $a$ to $b$ as the area of the rectangle $f(a)$ by $ab$.

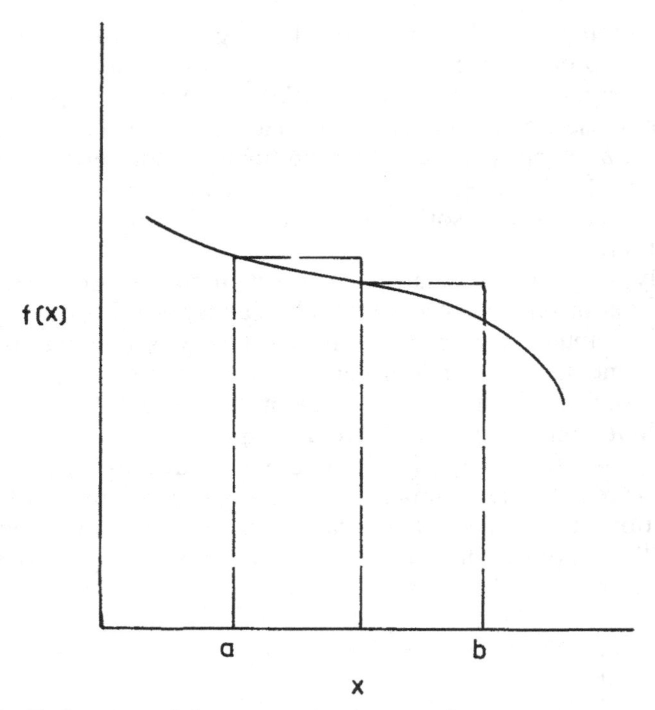


Fig. 7-7. Estimation of the area under $f(x)$ vs $x$ from $a$ to $b$ using two rectangles.

This is the strategic concept behind some of the most powerful computer methods known: *select an iterative method which, though it may give a very poor approximation to the true answer initially, may be made to approach the true answer on successive approximations*. Approach to the true answer is called *convergence*. The rectangle area method may be made to converge on the true answer by taking more rectangles, as seen in the sequence of Figs. 7-6, 7-7, and 7-8, and the last of which approximates the area under the curve as the sum of the areas of four rectangles.

## Program 7-2

This program applies the concept just stated to the standard normal curve, Eq. (7-22). Properties of convergence are investigated for 1, 10, 100, 1000, and 10,000 iterations in the program commentary. It should be mentioned in passing that the rectangle estimation method just described was chosen for its simplicity. There are more efficient methods, i.e., methods that converge on the true solution more rapidly. The advantage of more rapid convergence is that the difference between the approximate value and the true value may be brought to within a predetermined limit in fewer iterations. This in turn results in a saving of computer time which may be

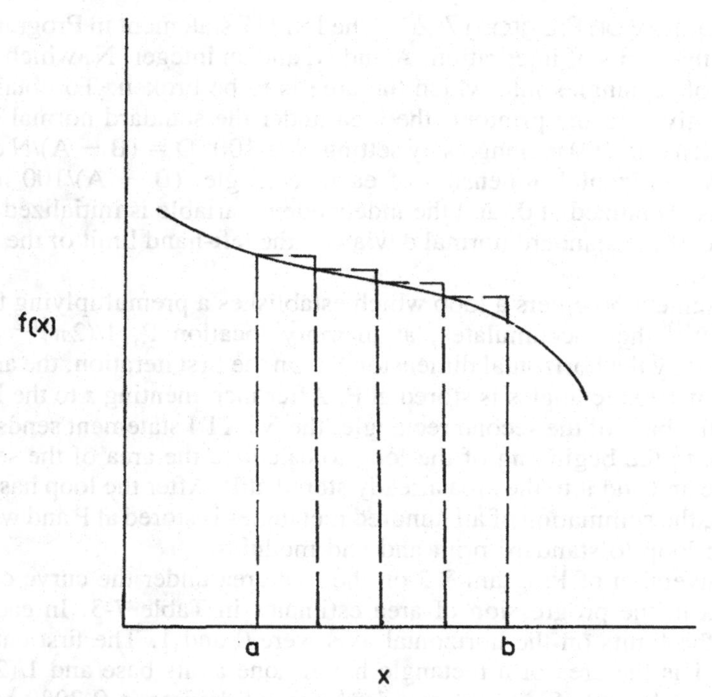


Fig. 7-8. Estimation of the area under $f(x)$ vs $x$ from $a$ to $b$ using four rectangles.

very expensive for a large machine. The interested reader is referred to methods based on the trapezoidal rule and Simpson's rule in Dence (1975). The general topic of numerical analysis is treated in detail in Acton (1970).

```
10  INPUT A,B,N
20  LET D=(B-A)/N
30  LET P=0
40  LET Z=A
50  FOR I=1 TO N
60  LET M=1./(2.*3.14159)**.5
70  LET P=P+M*(EXP(-Z*Z/2))*D
80  LET Z=Z+D
90  NEXT I
100 PRINT A,B,P
110 END

READY
RUNNH


?0,1,100
 0              1              0.342128


TIME:  0.27 SECS.
```

Commentary on Program 7-2. The INPUT statement in Program 7-2 reads in the limits of integration, A and B, and an integer, N, which is the number of rectangles into which the area is to be broken. To obtain the estimate given in the printout, the area under the standard normal curve was broken into 100 rectangles by setting N = 100. D = (B − A)/N determines the horizontal dimension of each rectangle, (B − A)/100 in this case; P is initialized at 0, and the independent variable is initialized at A, the value of the standard normal deviate at the left-hand limit of the interval.

Statement 50 enters a loop which establishes a premultiplying factor of $1/(2\pi)^{1/2}$ then accumulates, at memory location P, $1/(2\pi)^{1/2}\ e^{-z^2/2}$ multiplied by the horizontal dimension D. On the first iteration, the area of the first of 100 rectangles is stored at P. After incrementing $z$ to the lower limit of the base of the second rectangle, the NEXT I statement sends control back to the beginning of the loop to calculate the area of the second rectangle and add it to the area already stored at P. After the loop has been satisfied, the summation of all hundred rectangles is stored at P and we exit from the loop to standard print and end modules.

Conversion of Program 7-2 on the true area under the curve can be observed in the progression of area estimates in Table 7-3. In each instance, the limits on the horizontal axis were 0 and 1. The first entry in Table 7-1 is the area of a rectangle having one as its base and $1/(2\pi)^{1/2}$ $e^{-z^2/2}$ as its height. Since the standard normal deviate is 0.3989 at $z$ = 0, the area of the rectangle is also 0.3989, a poor approximation to the true value of 0.3413.

Increasing the number of iterations to ten increases the rectangles taken to a like number, resulting in a much better area approximation of 0.3489, one that is only 2.2% in error. Taking 100 rectangles yields 0.3421, the datum output by Program 7-2 above. One thousand iterations yield 0.3414, 0.029% in error, and 10,000 iterations yield 0.34135. One would not expect the summed area to go below the true integral because, by the method chosen, there should always be a tiny corner of each rectan-

Table 7-3
Area of Summed Rectangles for Program 7-2 (the True Value of the Integral is 0.3413.)

| Number of rectangles | Summed area |
|---|---|
| 1 | 0.3989 |
| 10 | 0.3489 |
| 100 | 0.3421 |
| 1000 | 0.3414 |
| 10,000 | 0.3413 |

gle above the curve of a monotonically decreasing function leading to a positive error. The last calculation in Table 7-2 takes about 1/2 h on a 1.78 MHz microcomputer.

## Tabulated Areas Under the Normal Curve

The area under the normal curve is so important in statistics that most books on the subject have tables of the area under the curve for intervals from $z = 0$ to $z = 3.0$. Use of the table to determine areas over different intervals is illustrated in the following exercise.

Table 7-4
Areas Under One Half of the Normal Curve

| z | .00 | .01 | .02 | .03 | .04 | .05 | .06 | .07 | .08 | .09 |
|---|---|---|---|---|---|---|---|---|---|---|
| 0 | 0 | 39 | 79 | 119 | 159 | 199 | 239 | 279 | 318 | 358 |
| 0.1 | 398 | 437 | 477 | 517 | 556 | 596 | 635 | 674 | 714 | 753 |
| 0.2 | 792 | 831 | 870 | 909 | 948 | 987 | 1025 | 1064 | 1102 | 1141 |
| 0.3 | 1179 | 1217 | 1255 | 1293 | 1330 | 1368 | 1405 | 1443 | 1480 | 1517 |
| 0.4 | 1554 | 1591 | 1627 | 1664 | 1700 | 1736 | 1772 | 1808 | 1844 | 1879 |
| 0.5 | 1914 | 1949 | 1984 | 2019 | 2053 | 2088 | 2122 | 2156 | 2190 | 2223 |
| 0.6 | 2257 | 2290 | 2323 | 2356 | 2389 | 2421 | 2453 | 2485 | 2517 | 2549 |
| 0.7 | 2580 | 2611 | 2642 | 2673 | 2703 | 2733 | 2763 | 2793 | 2823 | 2852 |
| 0.8 | 2881 | 2910 | 2939 | 2967 | 2995 | 3023 | 3051 | 3078 | 3105 | 3132 |
| 0.9 | 3159 | 3186 | 3212 | 3238 | 3264 | 3289 | 3315 | 3340 | 3364 | 3389 |
| 1 | 3413 | 3437 | 3461 | 3484 | 3508 | 3531 | 3554 | 3576 | 3599 | 3621 |
| 1.1 | 3643 | 3665 | 3686 | 3707 | 3728 | 3749 | 3769 | 3790 | 3810 | 3829 |
| 1.2 | 3849 | 3868 | 3887 | 3906 | 3925 | 3943 | 3961 | 3979 | 3997 | 4015 |
| 1.3 | 4032 | 4049 | 4066 | 4082 | 4099 | 4115 | 4131 | 4146 | 4162 | 4177 |
| 1.4 | 4192 | 4207 | 4222 | 4236 | 4251 | 4265 | 4279 | 4292 | 4306 | 4319 |
| 1.5 | 4332 | 4344 | 4357 | 4370 | 4382 | 4394 | 4406 | 4418 | 4429 | 4441 |
| 1.6 | 4452 | 4463 | 4474 | 4484 | 4495 | 4505 | 4515 | 4525 | 4535 | 4545 |
| 1.7 | 4554 | 4564 | 4573 | 4582 | 4591 | 4599 | 4608 | 4616 | 4625 | 4633 |
| 1.8 | 4641 | 4648 | 4656 | 4664 | 4671 | 4678 | 4686 | 4693 | 4699 | 4706 |
| 1.9 | 4713 | 4719 | 4726 | 4732 | 4738 | 4744 | 4750 | 4756 | 4761 | 4767 |
| 2. | 4772 | 4777 | 4783 | 4788 | 4793 | 4798 | 4803 | 4807 | 4812 | 4817 |
| 2.1 | 4821 | 4825 | 4830 | 4834 | 4838 | 4842 | 4846 | 4850 | 4853 | 4857 |
| 2.2 | 4861 | 4864 | 4868 | 4871 | 4874 | 4877 | 4881 | 4884 | 4887 | 4890 |
| 2.3 | 4893 | 4895 | 4898 | 4901 | 4903 | 4906 | 4908 | 4911 | 4913 | 4916 |
| 2.4 | 4918 | 4920 | 4922 | 4924 | 4926 | 4928 | 4930 | 4932 | 4934 | 4936 |
| 2.5 | 4937 | 4939 | 4941 | 4942 | 4944 | 4946 | 4947 | 4949 | 4950 | 4951 |
| 2.6 | 4953 | 4954 | 4955 | 4957 | 4958 | 4959 | 4960 | 4962 | 4963 | 4964 |
| 2.7 | 4965 | 4966 | 4967 | 4968 | 4969 | 4970 | 4971 | 4971 | 4972 | 4973 |
| 2.8 | 4974 | 4975 | 4975 | 4976 | 4977 | 4978 | 4978 | 4979 | 4980 | 4980 |
| 2.9 | 4981 | 4981 | 4982 | 4983 | 4983 | 4984 | 4984 | 4985 | 4985 | 4986 |

[a] Compute z to three digits, locate the first two digits in the column to the extreme left and select the entry corresponding to the third digit as you read across. Each entry is a four digit number (±0.0001) with a leading decimal point. Thus z = 1.87 leads to 0.4693 and z = 0.04 leads to 0.0159. Note that the top three rows are displaced to the left.

### Exercise 7-5

Determine the area under the standard normal curve and between the following limits on $z$

(a) 0.00 to 1.43
(b) 0.68 to 1.98
(c) −1.44 to 1.44
(d) −0.75 to 2.51

Solution 7-5. (a) The first two digits are located in the left-hand column in Table 7-2 and the third digit is located on the top row. The entry corresponding to the intersection of the row and column so located is 0.4236, which is the area under the curve. Since the curve is normalized, the area has no units.

(b) Determine the area from zero to the larger limit first, then subtract the area from zero to the smaller limit to find the desired area

$$0.4761 - 0.2517 = 0.2244$$

(c) Since the curve is symmetrical about zero, only one half of the areas need be tabulated. The area from 0 to +1.44 is the same as that from 0 to −1.44; hence

$$0.4251 + 0.4251 = 0.8502$$

(d) Using the same principle as we did in (c), we have

$$0.4939 + 0.2733 = 0.7672$$

## The Most Common Areas and Limits on *z*

A few values of $z$ and the corresponding areas under the normal curve are so important that they should be memorized. Table 7-4 shows the six most important ones. The discussion following it indicates a common use for each value.

Table 7-5
Values of $z$ and
Corresponding Areas
Under the Normal Curve

| $z$ | Area |
|---|---|
| ±1 | 0.6826 |
| ±2 | 0.9544 |
| ±3 | 0.9974 |
| ±1.645 | 0.9000 |
| ±1.96 | 0.9500 |
| ±2.575 | 0.9900 |

Experimental results are commonly given with the standard deviation from the mean as an indication of dispersion. If we have a series of results with a known standard deviation, we may expect future results to fall within $\pm\sigma$ 68.26% of the time, within $\pm 2\sigma$ 95.44% of the time, and within $\pm 3\sigma$ 99.74% of the time. Use is made of these rules in clinical laboratories by setting up a chart with analytical results of a known standard plotted horizontally as a function of the number of times the analysis is performed. Horizontal lines are also drawn at values $\sigma$, $2\sigma$, and $3\sigma$ above and below the known line. Typically, analysis would be performed on the standard each morning before any unknown samples are run. Results show a random scatter, but are not expected to fall outside the zone limited by $2\sigma$. When a value does fall outside this zone, a thorough checking procedure is instituted to find systematic errors in the procedure, such as gross operator error, spoiled reagent solutions, or instrument failure.

The preceding discussion leads to the concept of *confidence limits*. Going from the area under the curve in Fig. 7-5 back to the limits on $z$, we find that an individual measurement may be expected to fall within $\pm 1.645\sigma$ 90% of the time, $\pm 1.96\sigma$ 95% of the time, and $\pm 2.575\sigma$ 99% of the time. When, for example, analysis of a patient's bilirubin falls more than 1.96 standard deviations away from the mean of a large number of analyses of healthy individuals, we say that the individual is not a member of the healthy population, i.e., that something is wrong that is affecting that person's bilirubin level. When we make such a judgment on the grounds of $>1.96\sigma$ deviation from the mean, we expect to be right 95% of

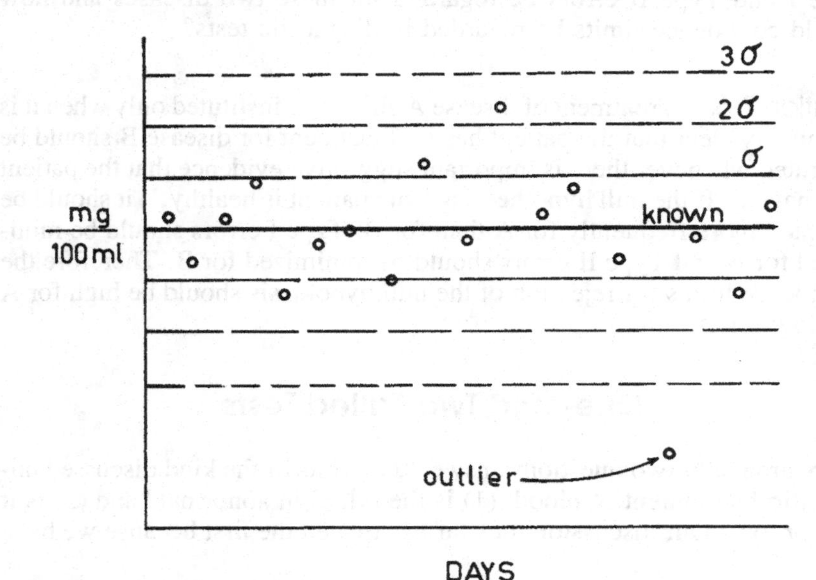


Fig. 7-9. Analytical calibration curve.

the time. We say that the judgment is made at the 95% or 0.95 confidence level. Values of $\sigma$ corresponding to 90, 95, and 99% confidence limits are given in Table 7-5. Note that there is no 100% confidence level in statistics.

If we say that the patient has an abnormal bilirubin analysis at the 95% confidence level, there is a 5% chance that we shall be wrong and that the patient is a normal patient with an unusually high bilirubin level (one patient out of 20).

The *null hypothesis* is usually the hypothesis that there is no significant difference between the measured value and the normal population. In the discussion above, we have rejected the null hypothesis and said that the patient is not a member of the normal population. If we reject the null hypothesis and we are wrong, we have committed a Type I error. If we fail to reject the null hypothesis and we are wrong, we have committed a Type II error. Selecting the confidence levels for a test depends on the relative seriousness of Type I and Type II errors. The lower the confidence level we require, the more likely will be Type I errors and *vice versa*. The more likely a Type I error, the less likely a Type II error.

### Exercise 7-6

Diseases A and B have the following characteristics: A has little chance of being fatal, and known procedures to cure it are expensive and unpleasant to the patient. Disease B is always or nearly always fatal. How should Type I and Type II errors be regarded for these two diseases and how should confidence limits be regarded in diagnostic tests?

Solution 7-6. Treatment of disease A should be instituted only when it is reasonably clear that the patient has it. Treatment for disease B should be instituted whenever there is important suggestive evidence that the patient may have it. If the null hypothesis is "the patient is healthy," it should be accepted more frequently for A than for B. Type I errors should be minimized for A and Type II errors should be minimized for B. Therefore the confidence limits for rejection of the null hypothesis should be high for A and low for B.

## One- and Two-Tailed Tests

There are really two questions answered by tests of the kind discussed under bilirubin content of blood: (1) is the bilirubin abnormal? and (2) is it high or low? Our discussion thus far has treated the first because we have considered measurements that are either higher than or lower than a specified multiple of $\sigma$. This is called a two-tailed test because it considers areas at both ends or tails of the normal curve.

Whether a test gives significantly high or significantly low results is a distinction of importance. Hyperbilirubinemia (high serum bilirubin) is more serious than hypobilirubinemia (low serum bilirubin); hence we might wish to consider only the positive end of the curve. This is a one-tailed test. If we rephrase the treatment above to say that an individual had a bilirubin count $>1.96\sigma$ *above* the mean, there is only one half or 2.5% probability that this observation could be the result of chance only. Thus, if a two-tailed test results in a 95% confidence level, a one-tailed test results in a 97.5% confidence level.

## Exercise 7-7

Suppose that the total serum cholesterol level in normal adults has been established as 200 mg/100 mL (mg%) with a standard deviation of 25 mg/100 mL. A patient's serum is analyzed for cholesterol and found to contain 265 mg/100 mL total cholesterol. (a) May we say at the 0.95 confidence level that the patient's cholesterol is abnormally elevated? (b) May we reach the same conclusion at the 0.99 confidence level? (c) What is the probability that the cholesterol reading obtained owed to chance factors alone?

Solution 7-7. This is a one-tailed test because the questions ask only about an abnormally high cholesterol level, not about a low one. Substituting into Eq. (7-21)

$$\frac{265 - 200}{25} = 2.6$$

The area under one half the normal curve is 0.4953, as seen in Table 7-2. Since this is a one-tailed test, we add the entire area on the low side of the mean, 0.5000, to obtain the total area, 0.9953. This tells us that 99.53% of the total serum cholesterol levels are expected to lie below the measured value for members of the normal population defined by $\bar{x}$ and $\sigma$. In other words, only $100 - 99.53 = 0.47\%$ of measured values are expected to fall higher than 265 mg/100 mL. The answers follow: (a) significant elevation at the 0.95 level, (b) significant elevation at the 0.99 level, and (c) 0.47%.

## Glossary

*Analytical Expression.* Representation of a functional relationship as a mathematical equation that may have a finite or an infinite number of terms.
*Confidence Level.* Probable ratio of right answers to the sum of right answers plus wrong answers.
*Convergence.* Approach of an approximate answer to the true answer.
*Gaussian Function.* Function of the form $e^{-z^2/2}$.
*Graphical Representation.* Representation of a functional relationship as a plot, usually two dimensional, of the dependent variable vs the independent variable, or one of the independent variables if there are more than one.
*Histogram.* Graph resulting from the representation of a frequency distribution as a series of contiguous rectangles with the height of the rectangle as the frequency and its base as the finite interval over which that frequency is observed.
*Integration.* Determination of the area under a curve between fixed limits on the independent variable.
*Normal Curve.* Curve describing the Gaussian function normalized to one.
*Normal Distribution.* Distribution of events, the frequency of which conforms to a Gaussian function.
*Normalization.* Mathematical procedure involving setting the integral of a function from minus to plus infinity equal to one.
*Null Hypotheses.* Any test hypothesis, but usually the hypothesis that observed deviations are random, i.e., owe to chance only.
*One-Tailed Test.* Directional test of the null hypothesis.
*Two-Tailed Test.* Nondirectional test of the null hypothesis.

## Problems

*1.* What is the area under the Gaussian curve between the following limits on the standard normal deviate?

(a) −0.60 and +0.60
(b) −0.15 and +1.87
(c) 0.145 and 1.855

Hint: use an interpolation method for (c).

*2.* Calculate the standard normal deviate for a measurement of 18.6 grams given a population mean of 20.0 grams and a standard deviation of 1.5 grams.

*3.* A large set of chemical analyses for the sulfur content of a given sulfide ore had %S = 31.44, with a standard deviation of 1.13%. An analysis was done on an ore sample which yielded 29.74% S. At what level of confidence may we say that the new sample was not taken from the original ore? If we make such a statement, what is the probability that we shall be wrong?

*4*. Suppose a measurement is found to deviate significantly from the population mean at the 0.90 confidence level, and we reject the null hypothesis that the measurement is a member of the population. Suppose further that we are wrong. What type error have we made? If we accept the null hypothesis and are wrong, what type error have we made?

*5*. Defining a healthy population as one that is normally distributed and has a total serum cholesterol of 200 mg/100 mL with a standard deviation of 25 mg/100 mL, may we say that a patient with a cholesterol count of 247 mg/100 mL has an elevated cholesterol count at the 0.95 confidence level? May we draw this conclusion at the 0.99 level?

*6*. Let us define a healthy population as one that is normally distributed and has a total serum cholesterol of 200 mg/100 mL with a standard deviation of 25 mg/100 mL. A patient's total serum cholesterol is analyzed and found to be 238 mg/100 mL. At what confidence level may we say that the patient's cholesterol level is abnormal?

*7*. What proportion of Type I errors relative to the total decisions made would you expect to make when applying a 0.95 confidence level criterion to random samples drawn from a normal population?

*8*. Modify Program 7-1 so as to graph such functions as $y = \sin x$, $y = x^2$, or any other functions that may interest you. You will have to use arbitrary multiplicative or additive factors to make the curve fit the page properly.

*9*. Modify Program 7-2 to integrate functions that you can check from the known closed solutions, e.g., $y = \int_0^2 x^2 dx$, $y = \int_1^3 (1/x)\, dx$, etc.

*10*. Show that Eq. 7-11 leads to Eq. 7-12.

*11*. Show that Eq. 7-14 leads to Eq. 7-15.

*12*. Write a program to reproduce the first horizontal line in Table 7-4, e.g., the ten probabilities from z = 0.0 to 0.9.

## Bibliography

F. S. Acton, *Numerical Methods That Work,* Harper and Row, New York, 1970.

L. N. Balaam, *Fundamentals of Biometry,* George Allen & Unwin, London, 1975.

J. B. Dence, *Mathematical Techniques in Chemistry,* Wiley-Interscience, New York, 1975.

J. H. Zar, *Biostatistical Analysis,* Prentice-Hall, Englewood Cliffs, New Jersey, 1974.

# Chapter 8

# Determining Probabilities

## How to Apply the Chi Square Test and the Student's *t*-Test

In general, real distributions are not the same as predicted or calculated distributions. A common and important problem in statistics is to determine how much deviation from a predicted distribution may be ascribed to pure chance. In other words, we would like to know whether the sample we have tested is really a legitimate representative of the infinite population for which the distribution was calculated. As usual in statistics, we cannot say with certainty that the sample is or is not a representative of the infinite population, or as is said, has been *drawn from* the infinite population. We can, however, use statistics to obtain the shrewdest possible guess and to calculate the probability of our being wrong. In cases where statistics is applicable, this is all that can be done. Statistics is the mathematics of events in the future or events for which some essential information has been obscured from us so that we cannot calculate an unequivocal answer.

As an example, consider a dice game in which we suspect that the dice are loaded. The owner of the dice is not likely to permit their careful inspection, particularly if he is cheating; hence information necessary for a definite decision whether or not the dice are loaded is not available. We wish to make a statistical decision here just as we did for single data points using the normal distribution. Here, however, we have a distribution of data points rather than a single datum or mean. We have already calculated the predicted distribution for the result of throwing two dice, which is redrawn for comparison as Fig. 8-1a.

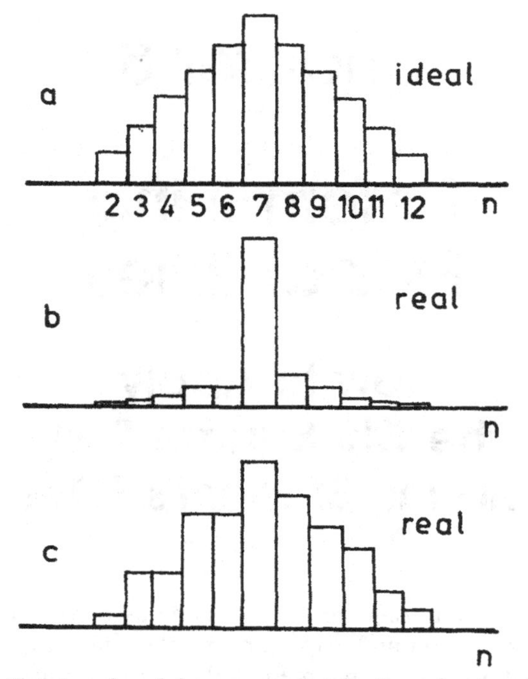


Fig. 8-1. Predicted and real frequency distributions for throwing two dice.

The real distribution shown in Fig. 8-1b is weighted very heavily in favor of 7, and common sense would lead us to decide that the dice are loaded. The number of throws is not given in the figure, but the facts that the distribution is smooth and that every possible throw is represented lead us to believe that the number of throws is large. Despite our qualitative decision that the dice are loaded, there is a finite (though small) probability that this distribution could be obtained purely by chance using honest dice. Hence there is a finite but small probability that our decision is wrong.

Figure 8-1c shows an intermediate case which is slanted in favor of 7 but not so much as case (b). The probability of obtaining this distribution by chance and with honest dice is considerably greater than in the previous case, and the likelihood that the decision "the dice are loaded" is wrong is correspondingly greater.

To put this kind of decision making on a quantitative statistical basis, let $F(n_i)$ be the real nonnormalized frequency of an event, in this case, the score $n_i$ from 2 to 12 inclusive.

Let $f(n_i)$ be the predicted normalized frequency for many trials, $N$. Ideally,

$$f(n_i) = F(n_i)\big/N \tag{8-1}$$

So that the predicted $F(n_i)$ for any sample is $Nf(n_i)$. The deviation of a sample frequency from the predicted frequency is

$$F(n_i) - Nf(n_i) \tag{8-2}$$

The relative deviation of a numerical result from the mean is

$$\mathrm{d} = (x - \bar{x})/\bar{x} \tag{8-3}$$

Analogously, the deviation of frequencies $d_{f(n_i)}$ of observation $n_i$ from the predicted frequency is

$$d_{F(n_i)} = [F(n_i) - Nf(n_i)]/Nf(n_i) \tag{8-4}$$

We wish to obtain a measure of deviation of one frequency distribution from another comparable to the variance hence we square the deviation to obtain

$$d^2_{F(n_i)} = [F(n_i) - Nf(n_i)]^2/Nf(n_i)^2 \tag{8-5}$$

Further, we would like to compare the entire real distribution with the predicted distribution over all events $n$, not just one, hence we sum over the squares of all possible deviations

$$\sum d^2_{F(n)} = \sum^{n} [F(n_i) - Nf(n_i)]^2/Nf(n_i)^2 \tag{8-6}$$

Here we invoke, somewhat arbitrarily, the principle of maximum likelihood, which states that a mean should be weighted so as to favor its more probable events. We can do this by multiplying by the predicted frequency, $Nf(n_i)$. The result is *defined* as $\chi^2$ (read "chi square")

$$\sum^{n} \{[F(n_i) - Nf(n_i)]^2/Nf(n_i)^2\} \, Nf(n_i) =$$

$$\sum^{n} [F(n_i) - Nf(n_i)]^2/Nf(n_i) \equiv \chi^2 \tag{8-7}$$

Calculated values of $\chi^2$ are a measure of agreement of an entire distribution with the corresponding predicted distribution. A table of values of $\chi^2$, degrees of freedom, and confidence levels is given as Table 8-1. If the value calculated for $\chi^2$ is greater than the tabulated value, the difference between the real frequency distribution and the predicted distribution is considered significant at the confidence level indicated at the top of the corresponding column. The number of degrees of freedom is the number of possibilities in the distribution minus the number of empirical constants in the predicted distribution (see below). This is because each empirical constant results from an equation connecting the variables. As previously

Table 8-1
Distribution of $\chi^2$

| | Probability of error | | | | | |
|---|---|---|---|---|---|---|
| *df* | 0.90 | 0.50 | 0.20 | 0.10 | 0.05 | 0.01 |
| 1 | .0158 | .455 | 1.642 | 2.706 | 3.841 | 6.634 |
| 2 | .211 | 1.386 | 3.219 | 4.605 | 5.991 | 9.210 |
| 3 | .584 | 2.366 | 4.642 | 6.251 | 7.815 | 11.345 |
| 4 | 1.064 | 3.357 | 5.989 | 7.779 | 9.488 | 13.277 |
| 5 | 1.610 | 4.351 | 7.289 | 9.236 | 11.070 | 15.086 |
| 6 | 2.204 | 5.348 | 8.558 | 10.645 | 12.592 | 16.812 |
| 7 | 2.833 | 6.346 | 9.803 | 12.017 | 14.067 | 18.475 |
| 8 | 3.490 | 7.344 | 11.030 | 13.362 | 15.507 | 20.090 |
| 9 | 4.168 | 8.343 | 12.242 | 14.684 | 16.919 | 21.666 |
| 10 | 4.865 | 9.342 | 13.442 | 15.987 | 18.307 | 23.209 |
| 11 | 5.578 | 10.341 | 14.631 | 17.275 | 19.675 | 24.725 |
| 12 | 6.304 | 11.340 | 15.812 | 18.549 | 21.026 | 26.217 |
| 13 | 7.042 | 12.340 | 16.985 | 19.812 | 22.362 | 27.688 |
| 14 | 7.790 | 13.339 | 18.151 | 21.064 | 23.685 | 29.141 |
| 15 | 8.547 | 14.339 | 19.311 | 22.307 | 24.996 | 30.578 |
| 16 | 9.312 | 15.338 | 20.465 | 23.542 | 26.296 | 32.000 |
| 17 | 10.085 | 16.338 | 21.615 | 24.769 | 27.587 | 33.409 |
| 18 | 10.865 | 17.338 | 22.760 | 25.989 | 28.869 | 34.805 |
| 19 | 11.651 | 18.338 | 23.900 | 27.204 | 30.144 | 36.191 |
| 20 | 12.443 | 19.337 | 25.038 | 28.412 | 31.410 | 37.566 |
| 21 | 13.240 | 20.337 | 26.171 | 29.615 | 32.671 | 38.932 |
| 22 | 14.041 | 21.337 | 27.301 | 30.813 | 33.924 | 40.289 |
| 23 | 14.848 | 22.337 | 28.429 | 32.007 | 35.172 | 41.638 |
| 24 | 15.659 | 23.337 | 29.553 | 33.196 | 36.415 | 42.980 |
| 25 | 16.473 | 24.337 | 30.675 | 34.382 | 37.652 | 44.314 |
| 26 | 17.292 | 25.336 | 31.797 | 35.563 | 38.885 | 45.642 |
| 27 | 18.114 | 26.336 | 32.912 | 46.741 | 40.113 | 46.963 |
| 28 | 18.939 | 27.336 | 34.027 | 37.916 | 41.337 | 48.278 |
| 29 | 19.768 | 28.336 | 35.139 | 39.087 | 42.557 | 49.588 |
| 30 | 20.599 | 29.336 | 36.250 | 40.256 | 43.773 | 50.892 |

discussed, any equation connecting the variables causes one variable to be dependent upon the remaining variables and reduces the number of degrees of freedom by one.

## Chi Square Detection of a Biased Coin

Application of the $\chi^2$ criterion to the simple problem of coin flipping might involve a test to determine whether a coin is *biased* or weighted—the simplest analogy to the question of loaded dice that opened this chapter. The problem is the same. Though we know that we shall not be able to say with certainty that the coin is or is not biased since, presumably, we are not able to make a minute physical investigation of it, we would like to infer an answer from observations of many tosses and to know the confidence limits within which that inference may be drawn.

Suppose the coin is tossed 1000 times and comes up 600 heads and 400 tails. This constitutes strong qualitative evidence that the coin is biased. Applying the $\chi^2$ test, for this real distribution of heads,

$$[F(n_1) - Nf(n_1)]^2 / Nf(n_1) = (600 - 500)^2/500$$
$$= (100)^2/500 = 10000/500 = 20$$

We must also look to the distribution of tails, which of necessity must appear simultaneously with the distribution of heads we have observed

$$[F(n_2) - Nf(n_2)]^2 / Nf(n_2) = (400 - 500)^2/500$$
$$= (-100)^2/500 = 10000/500 = 20$$

Chi square for the entire distribution is

$$\chi^2 = \sum [F(n_i) - Nf(n_i)]^2 / Nf(n_i) = 20 + 20 = 40$$

There are two possibilities in this distribution, heads and tails, but if we know the number of heads, we also know the number of tails because

$$n_H + n_T = 1000$$

hence the number of degrees of freedom is $2 - 1 = 1$. Looking into Table 8-1, we see that $\chi^2$ is much greater than 6.64, the entry for one degree of freedom at the $p = 0.01$ level. Therefore, our qualitative conclusion is confirmed; we may say that the coin is biased, with much less than 0.01 relative probability of error. Put in another way, we may conclude that the coin is biased with greater than 99% confidence that we are correct.

Suppose we had come to the same *ratio* of heads to tails, this time using only 1/10 as many trials as a basis for our conclusion, that is, in 100 trials, 60 heads and 40 tails were observed. Now,

$$\chi^2 = (60 - 50)^2 / 50 + (40 - 50)^2 / 50 = 100 / 50 + 100 / 50 = 4$$

At the $p = 0.01$ confidence level, 4 does not exceed the entry 6.64, hence we may not draw the conclusion that the coin is biased with this degree of confidence. The difference between the sample and the infinite population for which we have calculated the predicted frequencies is not significant at the 0.01 level.

Four is, however greater than the adjacent entry in Table 8-1, 3.85. The difference between distributions *is* significant at the $p = 0.05$ level. We may conclude that the coin is biased at slightly greater than the 95%

confidence limit with, however,slightly less than 5% chance of being wrong. A smaller number of trials gives us less confidence in our conclusion. Note that a probability of random fluctuation equal to or less than $p = 0.05$ leads to a relative confidence limit of 0.95 or 95%.

## Degrees of Freedom

Suppose we consider a situation with a greater number of possible events. We shall draw red, green, blue, or white balls from an urn. There are four possibilities but, for a given number of draws, say 100, once we know the number of R, G, and B balls drawn, we know that W = 100 − (R + G + B) hence the number of degrees of freedom is reduced by one. There are three degrees of freedom.

Suppose further that there are said to be 10, 20, 30, and 40% R, G, B, and W balls in the urn, and that we wish to take a sample to get the best estimate of whether this is true or not true. We draw a red ball, *replace it*, draw a green, with replacement, and so on. Since each ball is replaced after withdrawal, an infinite number of withdrawals can be made for any sample. This is the same as drawing without replacement from an infinite population. If, after 100 draws, we have drawn 10R, 15G, 30B, and 45W, we can calculate $\chi^2$ as follows

$$\chi^2 = (10 - 10)^2/10 + (15 - 20)^2/20 + (30 - 30)^2/30 + (45 - 40)^2/40 = 0 + 25/20 + 0 + 25/40 = 75/40 = 1.87$$

Looking at Table 8-1 again, at the 95% level, we have the entry 7.82 meaning that any value less than 7.82 cannot be said to indicate a nonrandom fluctuation at the 95% level. The calculated value for $\chi^2$ is less than this, hence we are not statistically justified in stating that the percentages given are wrong. In failing to reject the null hypothesis, "the contents of the urn are correctly represented," we are not necessarily accepting it, but only stating that we cannot reject it with 95% confidence.

## The Central Limit Theorem

Thus far, we have been restricted in most of our calculations to distributions that follow one of the model distributions: binomial, Poisson, or normal. Very many real distributions do not follow any of these models, but they may be treated statistically using a powerful theorem known as the *central limit theorem*. According to the central limit theorem, the mean of sample means approaches the population mean

$$\bar{x}_{\text{mean}} \rightarrow \mu_{\text{population}} \tag{8-8}$$

and the distribution of sample means about the population mean is a normal distribution *no matter how the population is distributed.* Thus we may always apply the methods of the last chapter to the distribution of a set of sample means about the grand mean.

Further, it can be shown that as the number of sample means becomes very large,

$$\sigma_{\bar{x}} = \sigma/(N)^{1/2} \tag{8-9}$$

where $\sigma_{\bar{x}}$ is the standard deviation of the sample means or the *standard deviation of the mean*, $\sigma$ is the standard deviation of the population, and $N$ is the number of measurements taken to calculate each sample mean. Generally, it is convenient to set up an experiment so that $N$ is the same for each sample mean, and we shall assume that this has been done in what follows.

## Exercise 8-1

Consider the normal red cell (erythrocyte) count to have been shown to have a standard deviation of $4.47 \times 10^5$ cells/mm$^3$. If blood samples are drawn from a group of individuals, divided into 20 aliquot portions each, and counted, what is the standard deviation of the sample means of each set of 20 counts? If only five aliquot portions are taken, what is the standard deviation of the mean?

Solution 8-1. The population standard deviation is assumed known. For 20 counts

$$\sigma_{\bar{x}} = 4.47 \times 10^5/(20)^{1/2} = 1.00 \times 10^5$$

For five counts

$$\sigma_{\bar{x}} = 4.47 \times 10^5/(5)^{1/2} = 2.00 \times 10^5$$

## Program 8-1

In the section on the standard normal deviate in Chapter 7 it was said that the Gaussian function broadens out with changes in $\sigma$. Indeed this was one of the reasons for defining $z$ in that section. Now that we have a plotting routine, we can investigate the influence of $\sigma$ on the Gaussian probability. Writing Eq. (7-20) as a probability and letting $\bar{x} = 0$ for convenience

$$p(x) = [1/(2\pi)^{1/2}\,\sigma] \exp -x^2/2\sigma^2 \tag{8-10}$$

A variation on Program 7-1 is given in which the Gaussian is not plotted in terms of the standard normal deviate, but in terms of the random variable,

$x$, itself. Thus the influence of the standard deviation is explicitly shown in the printout. The output of Program 8-1 shows the Gaussian for standard deviations of 0.7, 1.0, and 1.4.

```
10   Z=-3.
20   FOR I = 1 TO 61
21   S=1.
25   P=120/(SQR(2*3.14159)*S)
30   K=INT(P*EXP(-Z**2/(2*S**2)))
40   PRINT TAB(K)"*"
50   Z=Z+.1
60   NEXT I
70   END
```

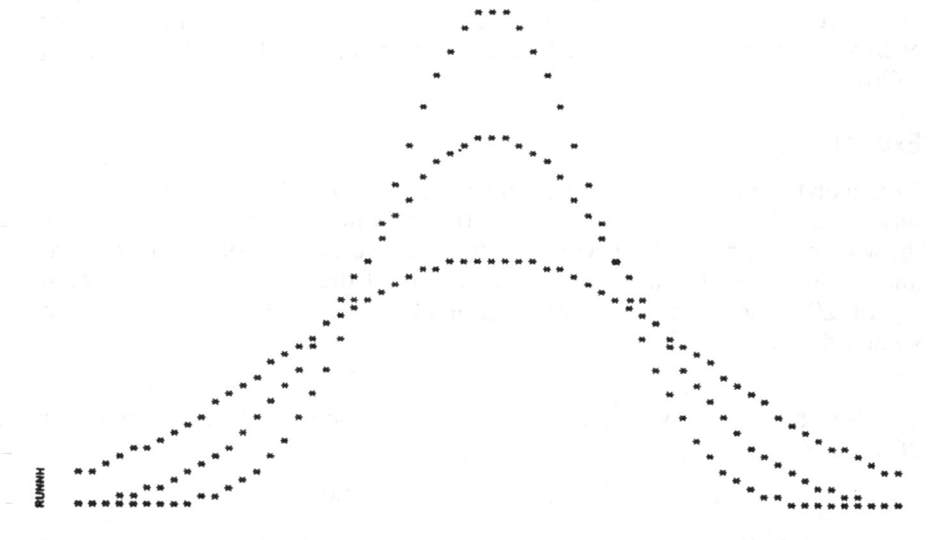

Commentary on Program 8-1. Statements 10 and 20 were taken directly from Program 7-1. Statement 21 gives a value to $S$, the standard deviation. This is the statement that is changed to investigate the influence of $S$ on the Gaussian distribution. Equation 8-10 has a premultiplier of $1/(2\pi)^{1/2}\sigma$ preceeding the exponential $\exp^{-x^2/2\sigma^2}$. These are translated into BASIC in statements 25 and 30 where the letter $S$ is used for $\sigma$. The remainder of the program is essentially the same as Program 7-1.

The output shows three curves superimposed on each other. This is done by manually rolling the paper back to one line before the RUN command, typing in a new statement 21 ($S = 0.7$, 1.0, 1.4, or other values), striking RUN over the previous RUN command whereupon the computer superimposes the new printout over the old one. There are better ways of superimposing one output on another without backing up the paper roll, but they add programming complexity.

The three curves shown give a graphic demonstration of the way a sharply-peaked Gaussian function broadens out as $\sigma$ is increased. In the process the peak must be lowered because the area under all *normalized* Gaussian curves must be the same.

## Exercise 8-2

Suppose that a very large study has been carried out and it has been shown that the normal red blood cell count in males between 20 and 35 years of age is $54.5 \times 10^5$ cells/mm$^3$ with a population standard deviation of $4.47 \times 10^5$ cells/mm$^3$. A further study has been instituted to determine whether a group of factory workers has an abnormal erythrocyte count. Twenty three individuals were tested and found to have a mean erythrocyte count of $50.6 \times 10^5$ cells/mm$^3$. We wish to know whether the difference is significant at the 0.95 and the 0.99 confidence levels.

Solution 8-2. Both $\sigma$ and $\mu$ are known. The method we employ is completely analogous to the one we employed when we wished to know whether the result of a single measurement was significantly different from the mean of a population that is known to be normally distributed.

The difference here is that we are not taking the population to be normally distributed, rather we are employing the central limit theorem by taking the distribution of sample means about the grand mean of sample means to be normally distributed. This is true whether the population is normally distributed or not.

The standard deviation of the mean is

$$\sigma_{\bar{x}} = \sigma/(23)^{1/2} = 9.32 \times 10^4 \text{ cells/mm}^3$$

We now wish to calculate a *z*-score, using an equation analogous to Eq. (7-36)

$$z = [\bar{x}_S - \mu]/\sigma_{\bar{x}} \tag{8-11}$$

where $\bar{x}_S$ indicates the sample mean, $\mu$ is the population mean, and $\sigma_{\bar{x}}$ has been defined

$$z = (50.6 \times 10^5 - 54.5 \times 10^5)/9.32 \times 10^4 = -4.18$$

We see that this *z*-score is more than four sigma units below the accepted mean, and without even consulting Table 7-2, we conclude that the set of results obtained is significantly different from the accepted mean at higher than both the 0.95 and the 0.99 level of confidence.

## Exercise 8-3

Suppose all the data in the previous exercise remained the same

$$\mu = 54.5 \times 10^5 \qquad \sigma = 4.47 \times 10^4$$
$$\bar{x}_S = 50.6 \times 10^5$$

with one exception, that we had studied only four blood samples. The question is the same: Is the difference significant at the 0.95 or 0.99 levels?

Solution 8-3. Recalculation shows

$$\sigma_{\bar{x}} = \sigma\big/(4)^{1/2} = 2.24 \times 10^5 \text{ cells/mm}^3$$

$$z = 3.9 \times 10^5\big/2.24 \times 10^5 = -1.74$$

Reference to Table 7-2 shows that 2(0.4591) or 0.9182/1.0 of the area under the Gaussian curve lies between $z$-scores of $-1.74$ and $+1.74$. If we were to make the statement that the difference between the mean of blood counts of the four selected factory workers is significantly different from the accepted mean, we would have a $100 - 91.82 = 8.18\%$ probability of committing a Type I error. The difference is not significant at either the 0.95 or the 0.99 confidence level.

This problem raises an interesting question: How is it that with exactly the same experimental data, we get opposite conclusions for the significance levels merely by reducing the number of subjects we observe? The answer lies in chance fluctuations and the convergence of sample means on the true population mean as we increase the number of observations. Here we see the operation of an intuitive notion that is probably agreed to by almost anyone who has done experimental work: results derived from large samples tend to be more reliable than those derived from small ones. This principle operates through the $N^{1/2}$ that appears in the expression for the standard deviation of the mean. As $N$ increases, the standard deviation of the mean becomes smaller, and the conditions set for ascribing the difference between an observed mean and a population mean to pure chance become more stringent.

## Exercise 8-4

Suppose we have reason to suspect depletion of red blood cells in our sample of four factory workers cited in the exercise above. We are not worried about whether their erythrocyte count is significantly higher than the accepted standard, only whether it is lower. Clearly, this is a directional question and leads to a one-tailed test in contrast to the two-tailed test used in the previous exercise.

Solution 8-4. Mathematical analysis of the data is identical to that given above leading to a $z$-score of $-1.74$ and an area under one-half of the Gaussian curve of 0.4591. Since we are not interested in the high side of the Gaussian curve, we ignore the cut-off at $+1.74$ and add 0.5000 to obtain 0.9591 as the total area under the curve. An area of 0.9591 contradicts the null hypothesis that the observed sample mean does not differ significantly from the accepted population mean at the 0.95 level, but not at the 0.99 confidence level. These last examples have been included to show that the same or similar measurements can lead to very different conclusions according to the criteria by which we evaluate them.

## Situations in Which the Population Mean Is Known But Not Its Standard Deviation

Not infrequently, the population mean is known or taken to be an established value, but there is no reliable value for $\sigma$, the population standard deviation. Statistical problems of the kind we have been solving arise, but they cannot be treated in exactly the same way because the lack of $\sigma$ precludes calculation of $\sigma_{\bar{x}}$ for a sample mean. The selection of a substitute value for $\sigma$ is a rather obvious one; we use the only thing we have, the standard deviation of individual measurements about the sample mean. This substitution has some important effects on the basic statistics that follow its use; in particular, it is almost certainly less reliable than the missing population mean would have been. Further, its unreliability is worse as the number of individual measurements that make it up becomes smaller.

Consequently, an entirely new statistic has been worked out for use in situations where the population standard deviation is unknown. It is called Student's $t$-statistic or simply the $t$-statistic. The methodology involved is not greatly different from that used in the $z$-test already discussed. The principal difference is in the use of a $t$-distribution in place of the $z$-distribution. The $t$-distribution curve resembles the Gaussian curve, but is somewhat broader, as might be supposed, because of the lower reliability of the sample mean when used as an approximation to the population mean. The broadness of the $t$-distribution results in acceptance of the null hypothesis at a given confidence level, where the same data might lead to its rejection if the $z$-distribution had been the applied criterion. This is because the $t$-distribution takes into account not only chance fluctuations of the sample mean from the population mean, but also chance fluctuations of the sample standard deviation away from the missing population standard deviation. There is simply more overall room for fluctuations owing to chance alone.

A second difference between the normal distribution and the $t$-distribution is that there is only one normal curve, while there is an entire family of $t$-curves according to the number of measurements in the sample mean. As might be supposed from the discussion above, broad, flat $t$-curves correspond to small samples. The larger the sample is, the closer its $t$-curve approaches a normal distribution.

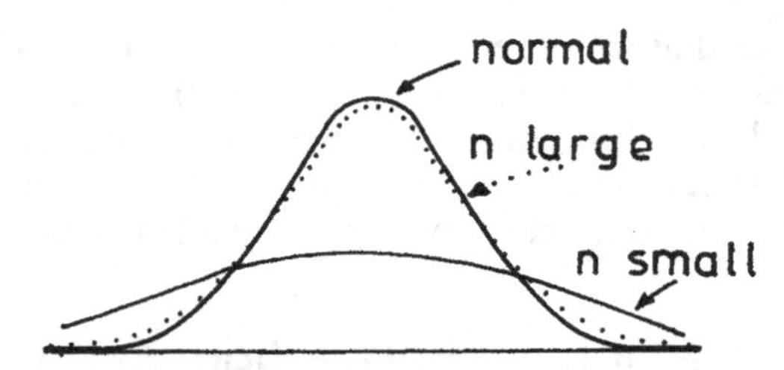


Fig. 8-2. Normal and Student's $t$-distribution for three and twenty degrees of freedom.

With Student's $t$-distribution, we once again encounter the question of degrees of freedom. Recall from earlier treatments that the number of independent algebraic variables constitutes the number of degrees of freedom. In general, the number of independent variables is the number of variables minus the number of equations connecting them. This can be a difficult point in some systems, but for the simple systems we shall consider, there will always be $N$ individual measurements and one equation for each sample mean calculated, $\bar{X} = \Sigma X_i / N$. The only examples we shall encounter are $(N - 1)$ for determination of significance and confidence limits on one experimental mean and $(N - 2)$ for use in comparing two independent sample means.

From this point on we shall use $\hat{\sigma}$ as the symbol for the sample standard deviation indicating that it is being used as an approximation to or *best estimator* of the population deviation. We proceed in a way analogous to that used in applying the $z$-test. First, the standard deviation of the mean from the sample mean is calculated with appropriate notation change

$$\hat{\sigma}_{\bar{x}} = \hat{\sigma} / N^{1/2} \tag{8-12}$$

then the analog to the $z$-score, which is called the *t-score*, is calculated

$$t = \bar{x} - \mu / \hat{\sigma}_{\bar{x}} \tag{8-13}$$

An appropriate null hypothesis is set up and accepted or rejected at a preselected confidence level according to the area under one or both tails of the appropriate $t$-curve, determined by the degrees of freedom of the system. Integration under $t$-curves is not fundamentally different from integration under the normal curve, but it is a little more involved, particularly because there are, strictly speaking, an infinity of them. This infinity of $t$-curves needn't be too alarming, however, because, aside from a few for the very lowest number of degrees of freedom, they look pretty much alike and resemble the $z$-curve very closely. Results of these area calculations are given in Table 8-2, Critical Values of Student's $t$-Distribution. The number of degrees of freedom, $\nu$, is listed in a vertical column at the left of the table and the ratio of the means that might be expected to *fall beyond*

Table 8-2
Student's *t*-Distribution

| $\nu$ | Probability of Error 0.9 | 0.5 | 0.2 | 0.1 | 0.05 | 0.02 | 0.01 |
|---|---|---|---|---|---|---|---|
| 1 | .158 | 1.000 | 3.078 | 6.314 | 12.706 | 31.821 | 63.657 |
| 2 | .142 | .816 | 1.886 | 2.920 | 4.303 | 6.965 | 9.925 |
| 3 | .137 | .765 | 1.638 | 2.353 | 3.182 | 4.541 | 5.841 |
| 4 | .134 | .741 | 1.533 | 2.132 | 2.776 | 3.747 | 4.604 |
| 5 | .132 | .727 | 1.476 | 2.015 | 2.571 | 3.365 | 4.032 |
| 6 | .131 | .718 | 1.440 | 1.943 | 2.447 | 3.143 | 3.707 |
| 7 | .130 | .711 | 1.415 | 1.895 | 2.365 | 2.998 | 3.499 |
| 8 | .130 | .706 | 1.397 | 1.860 | 2.306 | 2.896 | 3.355 |
| 9 | .129 | .703 | 1.383 | 1.833 | 2.262 | 2.821 | 3.250 |
| 10 | .129 | .700 | 1.372 | 1.812 | 2.228 | 2.764 | 3.169 |
| 11 | .129 | .697 | 1.363 | 1.796 | 2.201 | 2.718 | 3.106 |
| 12 | .128 | .695 | 1.356 | 1.782 | 2.179 | 2.681 | 3.055 |
| 13 | .128 | .694 | 1.350 | 1.771 | 2.160 | 2.650 | 3.012 |
| 14 | .128 | .692 | 1.345 | 1.761 | 2.145 | 2.624 | 2.977 |
| 15 | .128 | .691 | 1.341 | 1.753 | 2.131 | 2.602 | 2.947 |
| 16 | .128 | .690 | 1.337 | 1.746 | 2.120 | 2.583 | 2.921 |
| 17 | .128 | .689 | 1.333 | 1.740 | 2.110 | 2.567 | 2.898 |
| 18 | .127 | .688 | 1.330 | 1.734 | 2.101 | 2.552 | 2.878 |
| 19 | .127 | .688 | 1.328 | 1.729 | 2.093 | 2.539 | 2.861 |
| 20 | .127 | .687 | 1.325 | 1.725 | 2.086 | 2.528 | 2.845 |
| 21 | .127 | .686 | 1.323 | 1.721 | 2.080 | 2.518 | 2.831 |
| 22 | .127 | .686 | 1.321 | 1.717 | 2.074 | 2.508 | 2.819 |
| 23 | .127 | .685 | 1.319 | 1.714 | 2.069 | 2.500 | 2.807 |
| 24 | .127 | .685 | 1.318 | 1.711 | 2.064 | 2.492 | 2.797 |
| 25 | .127 | .684 | 1.316 | 1.708 | 2.060 | 2.485 | 2.787 |
| 26 | .127 | .684 | 1.315 | 1.706 | 2.056 | 2.479 | 2.779 |
| 27 | .127 | .684 | 1.314 | 1.703 | 2.052 | 2.473 | 2.771 |
| 28 | .127 | .683 | 1.313 | 1.701 | 2.048 | 2.467 | 2.763 |
| 29 | .127 | .683 | 1.311 | 1.699 | 2.045 | 2.462 | 2.756 |
| 30 | .127 | .683 | 1.310 | 1.697 | 2.042 | 2.457 | 2.750 |
| 40 | .126 | .681 | 1.303 | 1.684 | 2.021 | 2.423 | 2.704 |
| 60 | .126 | .679 | 1.296 | 1.671 | 2.000 | 2.390 | 2.660 |
| 120 | .126 | .677 | 1.289 | 1.658 | 1.980 | 2.358 | 2.617 |
| ∞ | .126 | .674 | 1.282 | 1.645 | 1.960 | 2.326 | 2.576 |

the critical value is listed at the top of the table above the horizontal line, denoted $\alpha$. To illustrate the method for five degrees of freedom, read down the leftmost column to $\nu = 5$. Reading across the table to the column under 0.05, we find the entry 2.571 at the intersection of the row and column specified. This means that a system for which $t$ has been calculated to be 2.571 has a 0.05 or 5% probability that the $t$-score results from chance alone. Normally, we would take this value of $t$ or greater to cause rejection of the null hypothesis at the 0.95 level, and conclude that the sample is not a member of the parent population with that level of confidence.

### Exercise 8-5

Normal cardiac output is widely accepted as 2.20 liters per minute per square meter of body area (L/min m$^2$). If a group of 14 patients has a sample mean cardiac output of 2.36 L/min m$^2$, with a sample standard deviation of 0.24 L/min m$^2$, is the difference from the general population significant?

Solution8-5. Only $\mu$ is given, not $\sigma$; hence we use the best estimator of $\sigma$

$$\hat{\sigma}_{\bar{x}} = \hat{\sigma}\Big/(N)^{1/2} = 0.24\Big/(14)^{1/2} = 0.064 \text{ L/min m}^2$$

$$t = (2.36 - 2.20)/0.064 = 0.16/0.064 = 2.49$$

$$\nu = 14 - 1 = 13$$

The difference is significant at the 0.95 level, but it is not significant at the 0.99 level.

## Determination of Confidence Limits When No Population Parameters Are Known

Although we have seen many instances, primarily clinical, involving comparison of sample statistics or individual measurements to one or both population parameters, $\mu$ and $\sigma$, many researchers are seldom, if ever, priveleged to know either one. Often the research scientist is measuring something that has never been measured before; if he or she knew the population parameters, he wouldn't have the slightest interest in measuring them again. This chapter concludes with two common situations faced by the experimental scientist: affixing confidence limits to a quantity that has been measured for the first time, and determining the significance or lack of it for the difference between two sets of observations made under different but comparable conditions.

If in the first situation, an experimenter makes a series of quantitative measurements, it is a simple matter to determine the mean and standard deviation for the sample. Further, it is easy to select an arbitrary value, not the sample mean, and to determine the probability that the experimental data were really drawn from a population having, as its mean, the value selected. This seemingly futile bit of guesswork has important implications in fixing confidence limits on experimental measurements.

### Exercise 8-6

A nutritionist wishes to know the amount of energy given off when the unsaturated fat, triolein, is hydrogenated to its saturated counterpart, tristearin. The experimenter obtains a sample mean of 90.83 kcal/mol and a standard deviation of 1.58 kcal/mol for sixteen determinations of the energy. Since these experiments have been carried out for the first time, the investigator has no population mean or standard deviation. Suppose he guesses, for no reason at all, a population mean of 91.00 kcal/mol. What is the probability that the difference between 90.83 and 91.00 owes to chance alone?

Solution 8-6. The problem is one in $t$-statistics. The estimated standard deviation of the mean is

$$\hat{\sigma}_{\bar{x}} = \hat{\sigma}\Big/(N)^{1/2} = 1.58\Big/(16)^{1/2} = 0.395 \text{ kcal/mol}$$

The $t$-score is

$$t = (90.83 - 91.00)\Big/0.395 = 0.430$$

At $\nu = 15$, Table 8-2 shows that the probability of a chance fluctuation of this size is well over 0.5, so that the experimental mean could quite comfortably have been drawn from a population having a mean of 91.00 kcal/mol.

## Confidence Limits

As it stands, Exercise 8-6 is a rather pointless calculation because, although 91.00 kcal/mol could have been the true population mean, an infinite number of other would-be population means might have been guessed, and calculations could have been carried out identical to those above, except that the $t$-scores would have been different.

We begin to see the point of the exercise when we work it in reverse. Choosing a one-tailed test, we ask, "What is the highest population mean from which the actual sample mean could have been drawn at the 0.95 confidence level?'' Looking down the column headed 0.10 (for two tails) to $\nu = 15$, we find $t = 1.753$. Since we are not interested in the upper tail, $-1.753$ is the $t$-score that just permits our observed sample mean to stay within the lower limit at the 0.95 confidence level. We are able to calculate the hypothetical population mean, which we shall designate $\bar{\mathbf{x}}$, and which constitutes the *upper 0.95 confidence limit* on our sample mean. The computation is easily carried out:

$$t = 1.753 = (90.83 - \bar{\mathbf{x}})/0.395$$
$$\bar{\mathbf{x}} - 90.83 = 0.6924$$
$$\bar{\mathbf{x}} = 91.52$$

The same calculation carried out for the lower 0.95 confidence limit yields 90.14 kcal/mol. Our conclusion is that the confidence limits on the sample mean are 90.14–91.52 or that the energy found is $90.83 \pm 0.69$ kcal/mol. Each of these two limits is a one-tailed 0.95 confidence limit on what the population mean is from which we have drawn our sample mean. The upper limit is that for which there is 95% probability that the population mean is below the limit and the lower limit yields a 95% probability that the population mean is above it. In other words, there is a 5% probability that the population mean is above the upper limit and an equal probability that it is below the lower limit. By applying the two limits simultaneously, we have only a 90% probability that the population mean is between them. This is easily remedied however by going back to a two-tailed test, one for the upper limit and one for the lower limit. Now the $t$-score is exactly as it appears in Table 8-2. The entry in the 0.05 column at $\nu = 15$ is 2.131 leading to $90.83 \pm 0.84$ kcal/mol for the sample mean with 0.95 confidence limits. In general, the confidence limits on a measurement of this kind are

$$\bar{x} \pm t\,\hat{\sigma}_{\bar{x}} \tag{8-14}$$

where $t$ is selected for a two-tailed test at the described confidence level.

## The Difference Between Means

The second major question facing experimentalists in almost any field concerns the decision whether observed differences between sample means are ''real'' or result from chance fluctuations only. To treat this question, we shall introduce notation indicating that there are two different means, $\bar{x}_1$ and $\bar{x}_2$. The standard deviations of the sample means are indicated as before.

Now, the standard deviation of the *difference* between the two means is related to the individual standard deviations by Eq. (8-15)

$$\sigma_{\bar{x}_1-\bar{x}_2} = (\sigma^2_{\bar{x}_1} + \sigma^2_{\bar{x}_2})^{1/2} \quad (8\text{-}15)$$

if the sample is normally distributed and if a $z$-score is applicable (that is, if we know $\sigma$ for the population). The $z$-score is determined with a slight notation change to show that we are using a standard deviation of the difference between means one and two:

$$z = (\bar{x}_1 - \bar{x}_2)/\sigma_{\bar{x}_1-\bar{x}_2} \quad (8\text{-}16)$$

Once the $z$-score is known, significance decisions are made and confidence levels are assigned as before.

Usually, one does not know $\sigma$, hence we follow Student's $t$ procedure by substituting the sample means for the population mean. Now,

$$\hat{\sigma}_{\bar{x}_1} = \hat{\sigma}_1/(N_1)^{1/2} \text{ and } \hat{\sigma}_{\bar{x}_2} = \hat{\sigma}_2/(N_2)^{1/2} \quad (8\text{-}17)$$

where $\hat{\sigma}_1$ and $\hat{\sigma}_2$ are the two sample standard deviations, $N_1$ and $N_2$ are the measurements in each set, and $\hat{\sigma}_{\bar{x}_1}$ and $\hat{\sigma}_{\bar{x}_2}$ are the standard deviations of the means. The standard deviations of the difference between means is

$$\hat{\sigma}_{\bar{x}_1-\bar{x}_2} = (\hat{\sigma}^2_{\bar{x}_1} + \hat{\sigma}^2_{\bar{x}_2})^{1/2} \quad (8\text{-}18)$$

and $t$ is

$$t = (\bar{x}_1 - \bar{x}_2)/\hat{\sigma}_{\bar{x}_1-\bar{x}_2} \quad (8\text{-}19)$$

## Exercise 8-7

Two groups of mice were randomly selected: one group of fourteen mice was chosen as the experimental group, and the other as the control group of twelve. The experimental group was subjected to a stress environment and the control group was not. Both groups were sacrificed and the weight of the adrenal gland was determined for each mouse in order to seek out any abnormal enlargment of the gland relative to stress and its resulting fear response in the fourteen animals in the experimental group.

The stress group had a mean adrenal weight of 5.06 milligrams (mg) with a sample standard deviation of 1.73 mg. The control group had a mean adrenal weight of 4.18, with a standard deviation of 1.78 mg. Is the difference significant at the 0.95 level?

Solution 8-7. For clarity in notation, let us use the subscript S to indicate the stress group and C to indicate the control group. It is also useful in problem solving to array the data in an orderly way:

$$\bar{x}_S = 5.06 \text{ mg} \qquad \bar{x}_C = 4.18 \text{ mg}$$
$$\sigma_S = 1.73 \qquad \sigma_C = 1.78$$
$$N_S = 14 \qquad N_C = 12$$

Since we are going to use a $t$-distribution, $\sigma_S$ and $\sigma_C$ will be denoted $\hat{\sigma}_S$ and $\hat{\sigma}_C$ in what follows. Further, the standard deviation of the mean for the stress group and control group will be called $\hat{\sigma}_{\bar{x}_S}$ and $\hat{\sigma}_{\bar{x}_C}$ respectively. Now,

$$\hat{\sigma}_{\bar{x}_S} = \hat{\sigma}_S / N_S^{1/2} = 1.73 / \sqrt{14} = 0.462$$

$$\hat{\sigma}_{\bar{x}_C} = \hat{\sigma}_C / N_C^{1/2} = 1.78 / \sqrt{12} = 0.514$$

The standard deviation of the difference between means is

$$\begin{aligned}\hat{\sigma}_{\bar{x}_S - \bar{x}_C} &= (\hat{\sigma}^2_{\bar{x}_S} + \hat{\sigma}^2_{\bar{x}_C})^{1/2} \\ &= (0.213 + 0.264)^{1/2} \\ &= 0.691\end{aligned}$$

The $t$-score is

$$t = (\bar{x}_S - \bar{x}_C) / \hat{\sigma}_{\bar{x}_S - \bar{x}_C} = 0.88 / 0.691 = 1.27$$

The number of degrees of freedom is the total number of observations minus the number of equations connecting them: *two* in this case, because we have used 14 members of the data set to calculate $\bar{x}_S$ (one connecting equation) and 12 members of the set to calculate $\bar{x}_C$ (the other connecting equation). Hence,

$$\nu = N_S + N_C - 2 = 24$$

For 24 degrees of freedom, the $t$-distribution resembles the $z$-distribution quite closely. The $t$-distribution gives, at 24 degrees of freedom, less than 0.90, 0.95 or 0.99 levels of significance between the means; we conclude that the adrenal mass was not a significant function of stress in this experiment.

## Glossary

*Bias.* A systematic distortion of a statistic or distribution.

*Central Limit Theorem.* Theorem stating that the grand mean of sample means is normally distributed regardless of the distribution of the sample means.

*Chi Square (or Chi Squared) Test.* Test for making a decision as to whether two distributions are statistically identical.

*Confidence Limits.* Limits on an observation within which the population mean is thought to lie with a predesignated probability.

*Degree of Freedom.* Independent variable.

*Distribution.* Collection of things placed in logical categories, specifically, probabilities of observing a variable as a function of the variable observed.

*Null Hypothesis.* Some hypothesis to be subjected to statistical tests, usually that there is no statistical difference between the observation or distribution studied and the population it is thought to represent.

*Standard Deviation of the Mean.* Standard deviation of sample means about their grand mean; it is normally distributed.
*Student.* Pseudonym for W. S. Gosset.
*Student's t-Test.* Statistical test replacing the normal or *z*-test when the population parameter $\sigma$ has been replaced by the sample statistic $s$.

## Problems

*1.* According to one theoretical prediction, a certain kind of tomato plant is supposed to produce three red tomatoes for each yellow one. In an experimental patch, 3739 red and 1063 yellow tomatoes were produced. Compute $\chi^2$ and decide whether the difference between the predicted and observed distributions is significant at the 99% level.

*2.* Rework the previous problem (based on Snedicor (1956)) but assume that the statistical sampling is much smaller, say one-tenth of the sampling previously specified. The hypothesis is still the same: three red to one yellow, but the sampling shows 374 red and 106 yellow. Calculate $\chi^2$ and estimate the level of significance between observed and theoretical.

*3.* A study has indicated that, among a normal population of healthy males, the mean body temperature, measured orally, is 98.38 degrees F with a standard deviation of 0.48°F. A group of 24 male students were innoculated and after one day, their temperatures were taken. The sample mean of the 24 students was 98.73. (1) Are the innoculated students significantly different in body temperature from the accepted population mean? (2) Do the innoculated students have a body temperature significantly higher than the population mean?

*4.* Suppose only one student had been innoculated and he had a temperature of 99.13 on the following day. Is this difference from the accepted mean significant (*a*) at the 95% level? (*b*) at the 99% level?

*5.* The data from Problem 3 remain the same, except that there are only eight students in the innoculated group. Calculate the significance and level of confidence of the observed mean for a one- and two-tailed test.

*6.* Suppose that a certain hospital has established 200 mg/100 mL as the mean value of serum cholesterol in normal patients. A study was done on eight business executives who have been subjected to stress and physical inactivity by the nature of their work. The group had a mean total serum cholesterol count of 220 mg/100 mL with a standard deviation of 40 mg/100 mL. Is this difference from accepted norm significant at the 0.95 level? Is there a significant elevation?

*7.* The highly unsaturated triglyceride, trilinolenin, has been found to have 270.86 kcal/mol more energy than its saturated counterpart, tristearin. The number of determinations contributing to this sample mean is 18 and the standard deviation of the individual determinations about their mean is 1.50 kcal/mol. What are the 0.95 confidence limits on this set of experimental measurements? What are the 0.99 confidence limits?

*8.* Two different analytical methods were used to determine the amount of prothrombin in plasma, lowered levels of which are associated with obstructive jaundice and liver malfunction. Both methods were used on the same pooled plasma source and ten analyses were run by each. The results were

$$\bar{x}_1 = 19.81 \qquad \bar{x}_2 = 19.54$$
$$S_1 = 1.44 \qquad S_2 = 1.81$$
$$N_1 = 10 \qquad N_2 = 10$$

Are the two methods statistically different at the 0.95 level?

*9.* Five students scored 71, 73, 67, 78, and 75 on an examination. Calculate the mean, standard deviation, and 95% confidence limits for this sample. If the assumption is made that the scores are randomly distributed about the mean, what is the probability that a randomly selected test score will be higher than 80? What is the probability that it will be lower than 60?

*10.* Ambiguity must be avoided in writing programs. In Program 8-1 if statement 30 were written 30 K = INT(P*EXP(−Z**2/2*S**2)), where is the ambiguity and what is the algebraic calculation that is made? (The calculation that we want and the one that is made in the actual program is $K = P\ exp^{-z^2/2\sigma^2}$.)

## Bibliography

H. D. Young, *Statistical Treatment of Experimental Data*, McGraw-Hill, New York, 1962.

F. E. Croxton, *Elementary Statistics With Applications in Medicine and the Biological Sciences*, Dover, New York, 1953.

F. J. Rohlf and R. R. Sokal, *Statistical Tables*, Freeman, San Francisco, 1969.

G. W. Snedicor, *Statistical Methods*, The Iowa State College Press, Ames, Iowa, 1956.

# Chapter 9

# Finding Linear Functions

## The Principle of Least Squares

Drawing a straight line by eye through a scattered set of experimental points, presumed to represent a linear function, is a subjective business at best. Moreover, no valid statistical parameters comparable to the standard deviation of the mean or well-defined confidence limits can be calculated for a line so drawn because your line will not necessarily come out exactly the same as my line. In this chapter, we shall use the principle of least squares to generate the equation of one and only one straight line for any given set of $xy$ data pairs. The line so obtained is the line that best fits the points subject to (1) the assumption of linearity and (2) the assumptions of the least squares method. We shall use the simplest case of a linear function passing through the origin to introduce the method and set up the ground rules. The more complicated case of a linear function not passing through the origin will be solved by a method that is general and will be extended to nonlinear functions in the next chapter and functions of many variables in the chapter after that.

The probability function governing an observation of outcome $x_i$ from among a continuous random distribution of possible events $x$ having a population mean $\mu$ and population standard deviation $\sigma$ is

$$p(x_i) \propto \exp \left\{ - (x_i - \mu)^2 \Big/ 2\sigma^2 \right\} \qquad (9\text{-}1)$$

The probability of observing events according to a certain *distribution* requires that event $x_1$ occur with frequency $f_1$, $x_2$ occur with frequency $f_2$, and so on. This is a problem in simultaneous probabilities,

$$p(x_1 \text{ and } x_2 \text{ and} \ldots) = p(x_1) \cdot p(x_2) \cdot \ldots = \prod^{N} p\,(x_i) \qquad (9\text{-}2)$$

where the capital pi $\prod$ indicates that we must take the product of all $N$ individual probabilities p

$$\prod p(x_i) \propto \prod \exp \{ -(x_i - \mu)^2 / 2\sigma^2 \} \tag{9-3}$$

Algebraically, when we multiply exponential numbers, we add exponents, e.g.,

$$e^a e^b = e^{a+b} = \exp (a + b) \tag{9-4}$$

hence

$$\prod^{N} p(x_i) \propto \prod^{N} \exp \{ -(x_i - \mu)^2 / 2\sigma^2 \}$$
$$\propto \exp (-\sum (x_i - \mu)^2 / 2\sigma^2) \tag{9-5}$$

Just as $e^{-x}$ takes its maximum value when $x$ takes its minimum possible value, the right side of Proportion 9-5 is a maximum when its exponent is a minimum. To minimize a fraction with a constant denominator, we minimize the numerator. Hence to minimize the exponent in Proportion 9-5, we need only set

$$\sum (x_i - \mu)^2 = \text{a minimum} \tag{9-6}$$

In so doing, we obtain the distribution having the maximum probability for the entire set of simultaneous events, i.e., the most probable distribution. Condition 9-6 amounts to selecting the value for $\mu$ so that the square of its difference from all the values of $x_i$, is as small as possible, not favoring or ignoring any of them. It is reasonable to suppose that, for Gaussian distributions, the "best" value for $\mu$ is the arithmetic mean, $\bar{x}$.

The principle described above, because it involves minimizing the sum of squares of deviations, is called the *principle of least squares*. This is the most powerful principle in elementary curve fitting and we shall use it repeatedly to solve increasingly difficult problems. In the more complicated cases, application of the principle of least squares entails very tedious calculations. Fortunately, the mathematics can be handled readily by computer and we shall deal with computer curve fitting in some detail.

If we accept the statement that, by selecting $\mu = \bar{x}$, Condition 9-6 is satisfied and the distribution of maximum probability is obtained, we can reverse the procedure and minimize the sum in Condition 9-6, thereby obtaining $\bar{x}$. This is an alternate way of calculating $\bar{x}$, as shown in Exercise 9-1.

## Exercise 9-1

For the data set $x_i = 2, 3, 7, 8, 10$ we have selected 5, 6, and 7 as values of $\mu$. Which value of $\mu$ is the best approximation to $\bar{x}$?

Solution 9-1. For the first value of $\mu$, $\mu = 5$,

$$(2 - \mu)^2 = (-3)^2 = 9$$
$$(3 - \mu)^2 = (-2)^2 = 4$$
$$(7 - \mu)^2 = (2)^2 = 4$$
$$(8 - \mu)^2 = (3)^2 = 9$$
$$(10 - \mu)^2 = (5)^2 = 25$$
$$\sum (x_i - \mu)^2 = 51$$

For $\mu = 6$,

$$\sum (x_i - \mu)^2 = 46$$

For $\mu = 7$,

$$\sum (x_i - \mu)^2 = 51$$

Therefore from among these three numbers, we select $\mu = 6$ as the best approximation to $\bar{x}$. In fact, the conventional calculation proves $\bar{x} = 6$ for this data set.

The process used above of selecting an arbitrary set of values for $\mu$ and finding which one leads to the least square deviation is not very efficient. Much better is to make use of the property that the first derivative of a function at its minimum is zero. It is also true that the first derivative of a function is zero at a maximum or an inflection point but these cases will not arise in our treatment.

If we set up $\Sigma(x_i - \mu)^2$ as a function of $\mu$, we obtain an infinite number of values of $\Sigma (x_i - \mu)^2$, one for each of an infinite number of possible values for $\mu$. All but one are not at the minimum, but we can find the minimum point by drawing a smooth curve through the points representing the function.

Better yet would be to take the first derivative of $\Sigma(x_i - \mu)^2$ with respect to $\mu$ and set it equal to zero. The reader will recall that the derivative of a dependent variable, in this case, $\Sigma(x_i - \mu)^2$ with respect to the independent variable is zero when the slope of the tangent to the curve is zero, i.e., when the tangent is a horizontal line.

The condition we wish to impose is

$$d/d\mu \sum(x_i - \mu)^2 = 0 \tag{9-7}$$

but the derivative of a sum is the sum of derivatives

$$d/d\mu \sum(x_i - \mu)^2 = \sum d/d\mu\ (x_i - \mu)^2 \tag{9-8}$$

and

$$d/d\mu\ (x_i - \mu)^2 = -2(x_i - \mu) \tag{9-9}$$

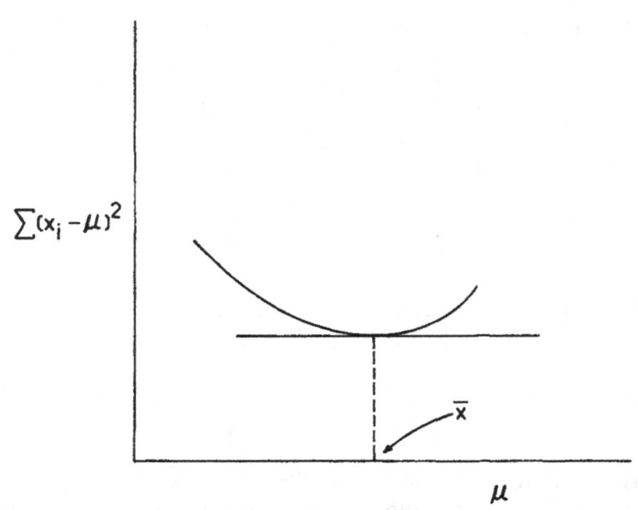


Fig. 9-1. The most probable value of $\mu$ as selected from the minimum in the curve expressing $\Sigma(x_i - \mu)^2$ as a function of $\mu$.

For the sum,

$$\sum d/d\mu\ (x_i - \mu)^2 = -\sum 2(x_i - \mu) = 0 \qquad (9\text{-}10)$$

which we have set equal to zero as the minimization condition. In this development, $\mu$ is called a *minimization parameter* because it is arbitrarily selected from an infinite number of possible choices so as to minimize the sum of squared deviations.

Let us rearrange the final result of Eq. (9-10).

$$2 \sum \mu = 2 \sum x_i \qquad (9\text{-}11)$$

We find that 2 may be dropped. Also, since $\mu$ is a constant and is summed over $N$ terms,

$$\sum_{i=1}^{N} \mu = N\mu \qquad (9\text{-}12)$$

hence

$$\mu = \sum x_i \Big/ N = \bar{x} \qquad (9\text{-}13)$$

which says that the least squares procedure leads to selection of $\bar{x}$ as the minimization parameter for a normally distributed sample. This is merely the conclusion that we reached on qualitative grounds earlier in this section. It is, of course, gratifying to find that we arrive, by rigorous application of this general principle, at the result $\mu = \bar{x}$, which seemed intuitively evident before.

## Linear Functions

Linear functions, as the name implies, lead to straight lines when they are plotted graphically. Linear functions of one independent variable may be written

$$y = f(x) \qquad (9\text{-}14)$$

where $x$ is taken as independent. This is arbitrary since the same function may be written

$$x = g(y) \qquad (9\text{-}15)$$

though this is the less conventional form. An example of a linear function in the form of Eq. (9-14) is obtained by multiplying $x$ by any constant,

$$y = mx \qquad (9\text{-}16)$$

where m may take any value, say 2 or 1/2 leading to the functions

$$y = 2x \qquad (9\text{-}17)$$

or

$$y = (1/2)\,x \qquad (9\text{-}18)$$

both of which are shown in Fig. 9-2. The slope, $m$, may also be zero, leading to a horizontal line, or negative, giving a downward line.

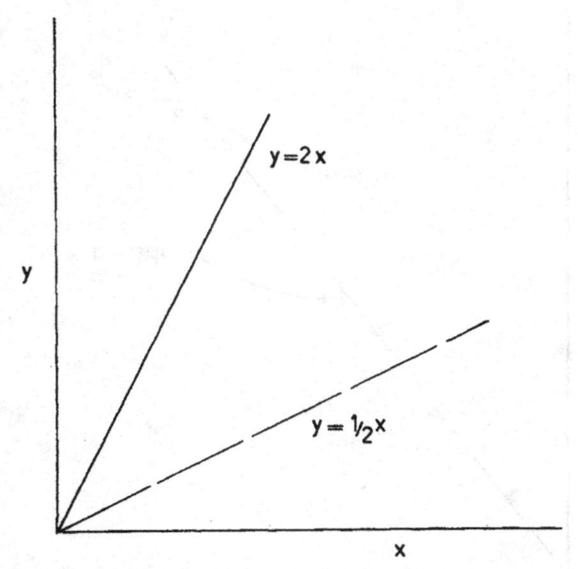


Fig. 9-2. Linear functions with $m = 2$ and $m = 1/2$.

Both of the functions pass through the origin, $y = x = 0$, which considerably simplifies their least squares treatment, hence functions of this kind will be considered first. A common example of linear functions that pass through the origin is that of the absorbance, $A$, of a solution as measured by a spectrophotometer at some wavelength

$$A = ac \tag{9-19}$$

Absorbance is a function of the concentration $c$ of the absorbing substance and has a slope a, called the *absorbancy*.

Figure 9-3 shows a linear functional relationship passing through the origin, just as does Fig. 9-2. However there is one important difference. Figure 9-2 shows no experimental points, implying that all points fall exactly on the line. This is certainly not the case for the real set of experimental variables, $c$, the known concentration, and $A$ the measured absorbance. Any real data set contains experimental errors or scatter, as shown in Fig. 9-3. Scatter may be very much less or very much more than the example illustrated. A straight line has been drawn through the points in Figure 9-3, in the way that is often done, by situating it by eye so that there are three points on the high side of the line and three points on the low side; we say that we have approximately "split the difference" resulting from experimental scatter and we suppose that this line is probably a better representation of the experimental data than a line drawn through any two specific points. This semiquantitative procedure, long followed in brief treatments

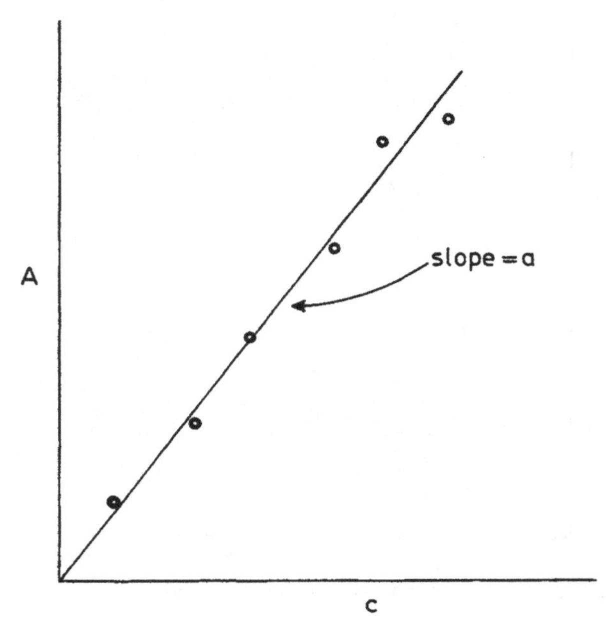


Fig. 9-3. Absorbance as a function of concentration for a simple solution.

of experimental data is valid, if approximate. The "best" straight line drawn through a simple set of points probably minimizes the deviations (and hence their squares) to a fairly good approximation. Thus the method of drawing the best curve through points is in accordance with the principle of least squares.

One person's "best" straight line may not be the same as another's however, particularly for large data sets, those with considerable scatter or data sets for which it is important to draw the best calculated result that the data can provide. With the power of the modern computer at our disposal, there is no longer any reason to settle for subjective approximations to the best line drawn through a set of points and journal editors are no longer content to accept data treated in this way.

## Least Squares Treatment of Linear Functions Passing Through the Origin

Although the example $A = ac$ was used to show that linear functions are commonly encountered in everyday experimental work, it is wise to work with the form, $y = mx$, to preserve generality and to emphasize that the method developed here can be applied to an indefinitely large number of different experimental situations that happen to have the same mathematical form. Indeed, that is the point of this book, that the methods developed here are general and may be used again and again in diverse experimental situations. The program resulting from this and the following curve-fitting methods should constitute an important part of the reader's program library. Sample applications are intended to illustrate the versatility of these fundamental programs.

If the linear function $y = mx$ were obeyed with perfect accuracy by the experimental data, we would always have

$$y/m - x = 0 \tag{9-20}$$

however this is never the case for real experimental data. Instead, each data point, $(x_1, y_1)$, $(x_2, y_2)$, . . . $(x_N, y_N)$, differs from the corresponding point on the line that best represents the function $y = mx$, the difference owing to experimental scatter. Consequently, Eq. (9-20) is not equal to zero; rather, we have a set of $N$ equations

$$\begin{aligned} d_1 &= y_1/m - x_1 \\ d_2 &= y_2/m - x_2 \\ &\vdots \\ d_N &= y_N/m - x_N \end{aligned} \tag{9-21}$$

represented by the general equation

$$d_i = y_i/m - x_i \tag{9-22}$$

or

$$md_i = y_i - mx_i \tag{9-23}$$

where the values of $d_i$ are deviations from the least squares straight line through the origin, and are not, in general, equal to zero.

We would like to minimize the sum of squares of the deviations, which can be done by taking the first derivative of Eq. (9-22) or (9-23) with respect to $m$ and setting it equal to zero

$$\frac{d}{dm} \sum (y_i - mx_i)^2 = 0 \tag{9-24}$$

or

$$\frac{d}{dm} \sum (y_i/m - x_i)^2 = 0 \tag{9-25}$$

The first of these derivatives is the simpler one to evaluate and leads to

$$m = \sum x_i y_i \Big/ \sum x_i^2 \tag{9-26}$$

## Exercise 9-2

Obtain Eq. (9-26) from Eq. (9-24).

### Solution 9-2.

$$\frac{d}{dm} \sum (y_i - mx_i)^2 = \sum \frac{d}{dm} (y_i - mx_i)^2 = \sum 2(y_i - mx_i)(-x_i)$$

Now

$$\sum(-2y_i x_i + 2mx_i^2) = -2 \sum (y_i x_i - mx_i^2) = 0$$

or

$$\sum y_i x_i - \sum mx_i^2 = \sum y_i x_i - m \sum x_i^2 = 0$$

whence

$$m = \sum y_i x_i \Big/ \sum x_i^2$$

Equation (9-25) leads to

$$m = \sum y_i^2 \Big/ \sum x_i y_i \tag{9-27}$$

which is an equivalent form of Eq. (9-26).

## Exercise 9-3

Equation (9-27) is not as simple to derive as Equation (9-26), but the method is the same. Derive Eq. (9-27) keeping in mind that

$$d(u/v) = (vdu - udv)\Big/v^2$$

Once one knows the slope of a linear function passing through the origin, one knows all that can be known about the function because linear functions are completely defined by the slope and any point on the curve, in this case, the origin.

## Exercise 9-4

Determine the slope of the least squares straight line passing through the origin and the following $x$, $y$ points

(1.0, 1.3), (2.0, 2.8), (3.0, 4.0), (4.0, 5.4), (5.0, 6.8)

Solution 9-4. By Eq. (9-26):

$$\sum x_i y_i = 74.5$$

$$\sum x_i^2 = 55.0$$

$$m = 74.5/55.0 = 1.35$$

## Program 9-1

Although application of BASIC to the problem of determining the least squares slope of a linear function passing through the origin does not bring in any new programming techniques, involving, as it does, nothing more than initializing and accumulating products and squares, there is more information that we might wish to extract from a set of experimental data. Merely knowing the slope through a set of experimental data points does not tell us anything about their dispersion, a piece of information that is certain to be very significant.

The following program and its modifications show how a simple program can be modified to be applicable to a number of increasingly difficult problems. The initial program in this series solves Eq. (9-26) for an arbitrary number of input $x$,$y$ pairs.

```
10   REM LEAST SQUARES FUNCTION THROUGH THE ORIGIN
20   DIM X(100),Y(100)
30   INPUT N
40   FOR I=1 TO N
```

```
50  INPUT X(I),Y(I)
60  LET S1=S1+X(I)*X(I)
70  LET S2=S2+X(I)*Y(I)
80  NEXT I
90  M=S2/S1
100 PRINT"    THE LEAST SQUARES SLOPE THROUGH THE ORIGIN IS",M
110 END

READY
RUNNH


 ?5
 ?1.31,1.74
 ?2.81,2.50
 ?3.11,3.21
 ?5.30,5.55
 ?7.11,8.00
    THE LEAST SQUARES SLOPE THROUGH THE ORIGIN IS        1.07818


TIME:  0.49 SECS.
```

Commentary on Program 9-1. After arbitrarily dimensioning $x$ and $y$ at 100, the actual number of data pairs in the set is requested by the INPUT N statement. At statement 40 control enters a FOR–NEXT loop that goes through N iterations requesting a data pair for $x$ and $y$ on each iteration and summing $x^2$ and $xy$ in locations S1 and S2. The slope is calculated as M = S2/S1 and printed out with a short alphabetical caption. The essence of this program is contained in the two accumulation steps, 60 and 70. We shall see more sophisticated use of accumulation loops in programs to come.

## Exercise 9-5

By changing two steps in Program 9-1 it is possible to calculate the slope by the alternate equation [Eq. (9-27)]. Do so for the same data set. The calculated slope is slightly different. This is because we have imposed passage through the origin on a function that does not exactly do so. If the data set were perfectly correlated and the function did indeed pass through the origin, the slope calculated by both methods would be the same. We shall treat correlation in the next section.

## Exercise 9-6

Prove algebraically that Eq. (9-26) equals Eq. (9-27).

## The Correlation Coefficient

Sometimes, it is not at all clear whether data are correlated or not. To test correlation, a *correlation coefficient* is calculated

$$r = \sum D_x D_y \Big/ \left(\sum D_x^2 \sum D_y^2\right)^{1/2} \qquad \text{(9-28)}$$

where $D_x = x_i - \bar{x}$ and $D_y = y_i - \bar{y}$.

From appropriate combination of Program 9-1 and its modifications, it is possible to write a program for calculating the correlation coefficient for an arbitrary number of data pairs.

Under the null hypothesis that the correlation is zero,

$$t = [r/(1 - r^2)^{1/2}](N - 2)^{1/2} \qquad \text{(9-29)}$$

We can calculate $r$ and $t$ for the question: Are the data correlated at, say, the 0.95 level? Comparison with Student's $t$-table enables us to make a decision whether the correlation is or is not significant at the preselected confidence limit in the way that we did for similar problems in Chapter 8.

### Exercise 9-7

Calculate the correlation coefficient and $t$-score for the data set used in Program 9-1. Is the correlation significant at the 0.99 level?

Solution 9-7. The correlation coefficient for this data set is given in the output of Program 9-2: $r = 0.99$.

$$\begin{aligned} t &= [r/(1-r^2)^{1/2}](N-2)^{1/2} \\ &= (0.99/0.141)(3)^{1/2} \\ &= 7.02(3)^{1/2} = 12.2 \end{aligned}$$

The quantity $N - 2$ appears because there are $N$ measurements (5 in this case) and two parameters in the linear equation $y = mx + b$, where $b = 0$ for this function. That is, there are 3 degrees of freedom for the set. The $t$ value for 99% confidence that the correlation of the set is *different* from zero is 12.924 (Table 8-2). The calculated value of $t$ is slightly less than this hence we conclude that correlation is significant at slightly less than the 0.99 confidence level.

### Program 9-2

```
10 DIM X(100),Y(100)
20 READ S1,S2,S3
30 DATA 0,0,0
35 PRINT "ENTER THE NUMBER OF DATA PAIRS"
```

```
40 INPUT N
45 PRINT "ENTER X FIRST, THEN Y; REPEAT"
50 FOR I=1 TO N
60 INPUT X(I),Y(I)
70 S1=S1+X(I)
80 S2=S2+Y(I)
90 NEXT I
100 M1=S1/N
110 M2=S2/N
120 FOR I=1 TO N
130 D1=D1+(X(I)-M1)*(Y(I)-M2)
140 D3=D3+(X(I)-M1)**2
150 D4=D4+(Y(I)-M2)**2
160 NEXT I
170 R=D1/SQRT(D3*D4)
180 PRINT "THE CORRELATION COEFFICIENT IS"R
190 END

READY
RUNNH


ENTER THE NUMBER OF DATA PAIRS
 ?5
ENTER X FIRST, THEN Y; REPEAT
 ?1.31,1.74
 ?2.81,2.50
 ?3.11,3.21
 ?5.30,5.55
 ?7.11,8.00
THE CORRELATION COEFFICIENT IS 0.990059


TIME:  0.23 SECS.
```

### Exercise 9-8

Write a commentary on Program 9-2.

## Applications

One application of the procedures just described is in comparing two methods of analysis, or comparing analytical results obtained by two different laboratory workers. If both methods or both technicians are working on the same samples, even though we should not expect all results to be identical because of random experimental scatter, we should expect the slope of ordered pairs of results by both methods or workers to be essentially one. A second common application has already been mentioned: applying Beer's law for a spectrophotometric investigation of a substance that does not contain any interfering substances absorbing light at the wavelength used. Beer's law states that the absorbance of light is a linear function of concentration for a specific wavelength of light and a fixed light path. In this case, of course, there is no reason to suppose that the slope will be one, indeed, it almost certainly will not be.

Frequently, we merely want to answer the question: Are factors $x$ and $y$ correlated? The answer may be that they are positively correlated, i.e., that for a high value of $x$, we may expect a high value of $y$, and vice versa. In this case, the correlation coefficient is positive. If $x$ and $y$ are perfectly and positively correlated, an exact linear equation connects them and knowledge of one permits exact calculation of the other. In this case, $r = +1.0$. Conversely, $x$ and $y$ may be perfectly negatively correlated ($r = -1.0$). A high value of $x$ coincides with a low value of $y$, and vice versa. An infinity of intermediate cases exists. No correlation at all, i.e., independence of $x$ and $y$ has $r = 0.0$. These applications will be illustrated in the following three examples.

## Exercise 9-9

A new laboratory technician has been hired to run determinations of sodium in blood serum by a flame photometric technique. The technician is given a series of standard solutions containing 0.20, 0.40, 0.60, 0.80, and 1.00 milliequivalents (meq) of sodium chloride per liter of water, and assays them to contain 0.19, 0.39, 0.62, 0.78, and 1.01 meq/L. What is the slope of the line of meq found versus meq taken? What are the units of the slope you have obtained? What is the correlation coefficient between meq found and meq taken?

Solution 9-9. The slope is 1.00. The units of the concentrations involved are meq/L/meq/L, which cancel, leaving a unitless number. The correlation coefficient is 0.9988, rounded to the proper number of significant figures, 1.00.

## Exercise 9-10

An excess of porphobilinogen in the urine is associated with hepatic disorders and lead poisoning. Porphobilinogen can be separated from other porphyrins by ion exchange chromatography and treated with *p*-dimethylaminobenzaldehyde to produce a red compound that absorbs light strongly at 550 nanometers (nm). A set of standard solutions was made up with concentrations of 50.0, 75.0, 100.0, 125.0, 150.0, 175.0, 200.0, 225.0, and 250.0 g/100 mL of porphobilinogen. Their absorbances $A$ after treatment were 0.040, 0.060, 0.088, 0.108, 0.120, 0.161, 0.179, 0.192, and 0.215. What is the slope of the line of A vs concentration in the original standards as obtained by this method? What are the units of slope? Three urine specimens were treated by this method and yielded absorbancies of 0.165,

0.178, 0.210. What were the porphobilinogen analyses on these three urine samples?

Solution 9-10. The slope of the Beer's law plot through the origin is

$$\text{Slope} = A/bc = 1.15 \times 10^3$$
$$\text{Units} = 100 \text{ mL/cm-mg}$$

because $A$ is a unitless number, $b$ is in cm, and $c$ is mg/100 mL. Specimens 1: 190, 2: 205, 3: 242 mg/100 mL.

### Exercise 9-11

Go back to the table of glucose infused versus glucose retained for 18 hospital patients (Table 2-2). What is the correlation coefficient of glucose infused and glucose retained? Is this correlation significant at the 0.95 level?

Solution 9-11. Program 9-3 yields

$$r = 0.965$$
$$t = [0.965/(0.0688)^{1/2}](16)^{1/2} = 14.7$$

The $t$ criterion of significance of *difference* from zero correlation for 16 degrees of freedom is 2.12. The data show significant linear correlation at the 0.95 level. Consulting a more comprehensive $t$ table, we find at 16 degrees of freedom that 14.7 far exceeds the entry for significance at the 0.999 confidence level. Thus, if we make the statement "glucose infusion is positively correlated with glucose retention," the probability of being wrong is far less than 0.1%.

## Least Squares with Two Minimization Parameters

Frequently, curves drawn through a set of experimental points do not pass through the origin. The standard form for a linear equation not passing through the origin is

$$y = mx + b \tag{9-31}$$

where $y$ is the dependent variable, $x$ is the independent variable, $m$ is the slope, and $b$ is the point at which the curve cuts the $y$ axis, called the $y$ intercept.

If there were no experimental error, Eq. (9-31) would apply perfectly, but since there is always experimental error, the data set produces a set of equations

$$y_1 = mx_1 + b + d_1$$
$$y_2 = mx_2 + b + d_2$$
$$\cdot$$
$$\cdot$$
$$\cdot$$
$$y_N = mx_N + b + d_N \tag{9-32}$$

where no experimentally determined $y$ is exactly the value of $y$ from the best straight line. Graphically, experimental points allow many straight lines to be drawn, each one satisfying exactly two points.

None of the lines in Fig. 9-4 is really suitable and most of them are clearly unsuitable. The curve that most of us would draw, which splits experimental scatter pretty well, would be close to curve (a). A curve that perfectly satisfies two experimental points, but is clearly inappropriate is curve (b). We would like to have a nonsubjective method of deciding which of the infinite number of lines that can be drawn through the points is the best. The problem is just as it was when we attempted to find a best line through the points representing a curve through the origin and the methods are identical. The result is, however, considerably more tedious

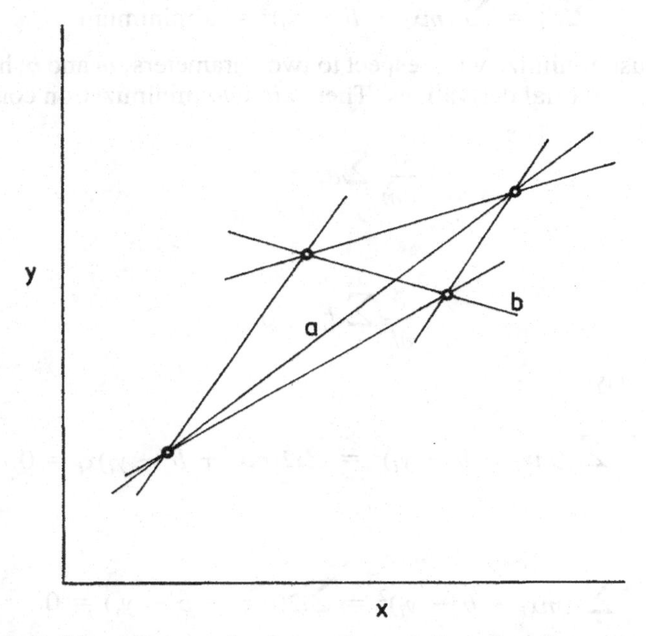


Fig. 9-4. Lines that can be drawn through four experimental points.

in its calculation than the previous case and in the process of finding it, we shall begin our consideration of simultaneous equations.

Equations (9-32) lead to a set of deviations from the best straight line representing them,

$$d_1 = (mx_1 + b) - y_1$$
$$d_2 = (mx_2 + b) - y_2$$
$$\vdots$$
$$d_N = (mx_N + b) - y_N \tag{9-33}$$

with the general form

$$d_i = (mx_i + b) - y_i \tag{9-34}$$

Each of the values, $(mx_i - b)$, tells us where $y_i$ "should be" according to the line and each $y_i$ tells us where the experimental value actually is.

If the deviations, $d_i$, are distributed in a random or Gaussian way about the best line representing the points, we may determine the equation of the best straight line by minimizing the sum of squared deviations

$$\sum d_i^2 = \sum (mx_i + b - y_i)^2 = \text{a minimum} \tag{9-35}$$

Now we must minimize with respect to two parameters, $m$ and $b$; hence we must take two partial derivatives. There are two minimization conditions,

$$\frac{\partial}{\partial m} \sum d_i^2 = 0 \tag{9-36}$$

and

$$\frac{\partial}{\partial b} \sum d_i^2 = 0 \tag{9-37}$$

which is to say

$$\frac{\partial}{\partial m} \sum (mx_i + b - y_i)^2 = \sum 2(mx_i + b - y_i)x_i = 0 \tag{9-38}$$

and

$$\frac{\partial}{\partial b} \sum (mx_i + b - y_i)^2 = \sum 2(mx_i + b - y_i) = 0 \tag{9-39}$$

The former equation reduces to

$$\sum mx_i^2 + \sum bx_i - \sum x_iy_i = 0 \tag{9-40}$$

and the latter to

$$\sum mx_i + \sum b - \sum y_i = 0 \tag{9-41}$$

These lead to

$$\sum x_iy_i = m\sum x_i^2 + b\sum x_i \tag{9-42}$$

and

$$\sum y_i = m\sum x_i + Nb \tag{9-43}$$

which are called the *normal equations* for this curve-fitting problem. The normal equations can be solved simultaneously to obtain

$$m = [N\sum x_iy_i - (\sum x_i)(\sum y_i)] \Big/ [N\sum x_i^2 - (\sum x_i)^2] \tag{9-44}$$

and

$$b = [(\sum y_i)(\sum x_i^2) - (\sum x_iy_i)(\sum x_i)] \Big/ [N\sum x_i^2 - (\sum x_i)^2] \tag{9-45}$$

In Eq. (9-36) and (9-37), partial notation has been introduced to account for our having to deal with derivatives of two variables rather than one, each taken with the assumption that the other variable is constant. We saw this method used in the section on propagation of errors.

## Program 9-3

Once one has the equations for slope and intercept of the least squares linear function passing through an arbitrary number of data points, one can write a program to solve them merely using elaborations on the initialization, accumulation, and arithmetic modules already shown. Using the variables $m$ and $b$, one can calculate the value of what $y$ "should be" for each value of $x$. Comparison of the true value of $y_i$ with the values of $y_i$ calculated from the linear least squares line leaves a *residual* that we called $d_i$ in Eq. (9-34). Program 9-3 performs all these functions and prints out the slope, intercept, input X and Y, the residuals under column heading R, and the least squares calculated values of $y_i$ under the column heading YCAL.

In addition, having the residuals, which are to a straight line what a deviation is to a point, we may obtain the standard deviation of the residuals in $y$ about the least squares straight line

$$s = \left[\sum R^2 \Big/ (N-2)\right]^{1/2} \tag{9-46}$$

where the number of degrees of freedom $N$ is reduced by two, one for the slope and one for the intercept.

```
1   REM LEAST SQUARES FIT TO A FIRST DEGREE EQUATION
10  DIM R(100),X(100)
11  DIM Y(100),D(100)
20  INPUT N
30  IF N<=0 GO TO 260
40  FOR I=1 TO N
50  INPUT X(I),Y(I)
60  NEXT I
70  READ S1,S2,S3,S4,S5
71  DATA 0,0,0,0,0
80  FOR I=1 TO N
91  S1=S1+X(I)
92  S2=S2+Y(I)
93  S3=S3+X(I)**2
94  S4=S4+X(I)*Y(I)
100 NEXT I
110 E=N*S3-S1**2
111 M=(N*S4-S1*S2)/E
112 B=(S2*S3-S1*S4)/E
120 FOR I=1 TO N
130 D(I)=M*X(I)+B
131 R(I)=Y(I)-D(I)
132 S5=S5+R(I)**2
140 NEXT I
150 C=N-1
151 G=SQR(S5/C)
152 X1=-B/M
160 PRINT N"POINTS, FIT WITH STD DEV "G
170 PRINT "SLOPE = "M",X INTERCEPT = "X1", Y INTERCEPT ="B
180 PRINT
190 PRINT "                X                  Y                  R                  YCAL
C"
200 PRINT
210 FOR I=1 TO N
220 PRINTUSING 230,X(I),Y(I),R(I),D(I)
230 :           ###.###          ###.###          ###.###          ###.###
231 NEXT I
240 PRINT
250 GO TO 20
260 PRINT
270  END

READY

RUNNH


 ? 11
 ? 0.,2.
 ? 10.,2.6
 ? 20.,3.55
 ? 30.,4.29
 ? 40.,4.85
 ? 50.,5.62
 ? 60.,6.25
 ? 70.,7.14
 ? 80.,7.35
 ? 90.,8.55
 ? 100.,9.26
 11 POINTS, FIT WITH STD DEV  0.152035
SLOPE =  7.14545E-2 ,X INTERCEPT = -28.1934 , Y INTERCEPT = 2.01455

            X                  Y                  R                  YCALC

          0.000              2.000             -0.015              2.015
         10.000              2.600             -0.129              2.729
         20.000              3.550              0.106              3.444
```

| X | Y | R | YCALC |
|---|---|---|---|
| 30.000 | 4.290 | 0.132 | 4.158 |
| 40.000 | 4.850 | -0.023 | 4.873 |
| 50.000 | 5.620 | 0.033 | 5.587 |
| 60.000 | 6.250 | -0.052 | 6.302 |
| 70.000 | 7.140 | 0.124 | 7.016 |
| 80.000 | 7.350 | -0.381 | 7.731 |
| 90.000 | 8.550 | 0.105 | 8.445 |
| 100.000 | 9.260 | 0.100 | 9.160 |

We may transfer what we have already learned about the Gaussian function to the normally distributed residuals about a straight line; approximately 68% of the points will lie within $\sigma$ of the line, 95% within 2 $\sigma$, and 99% within 3 $\sigma$ of the line. These ideas are shown graphically in Fig. 9-5 in which the normal curve (Gaussian) is drawn so that it projects out of the plane of the paper toward the reader. A more advanced treatment (Chatterjee, 1977) shows that confidence limits can be calculated for the slope and intercept of the least squares regression line just as they were for measurement of a single variable. Correlation coefficients can also be calculated for a linear function that does not pass through the origin.

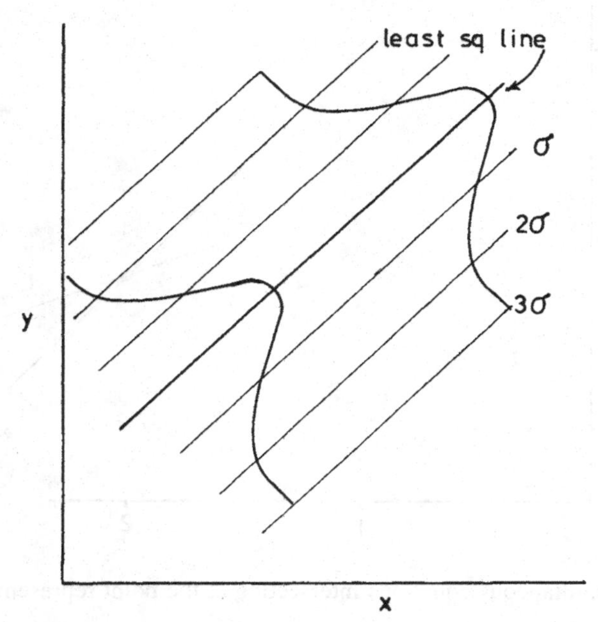


Fig. 9-5. Normally distributed residuals about a least squares straight line.

## Exercise 9-12

Write the commentary for Program 9-3.

## Simultaneous Equations

There are a number of ways of solving simultaneous equations, three of which will be considered in this book. A simple method is by Cramer's rule. To illustrate this method, we shall set up the equations in the most favorable way and introduce some of the nomenclature of matrix algebra.

We should recall simple simultaneous equations from elementary algebra, for example,

$$3x_1 + x_2 = 5$$
$$x_1 + x_2 = 3 \tag{9-47}$$

which have the solution set $x_1 = 1$ and $x_2 = 2$. The solution set is sometimes written as a number pair (1, 2), which is appropriate because it is a *point* of intersection of the line representing the first equation with that representing the second.

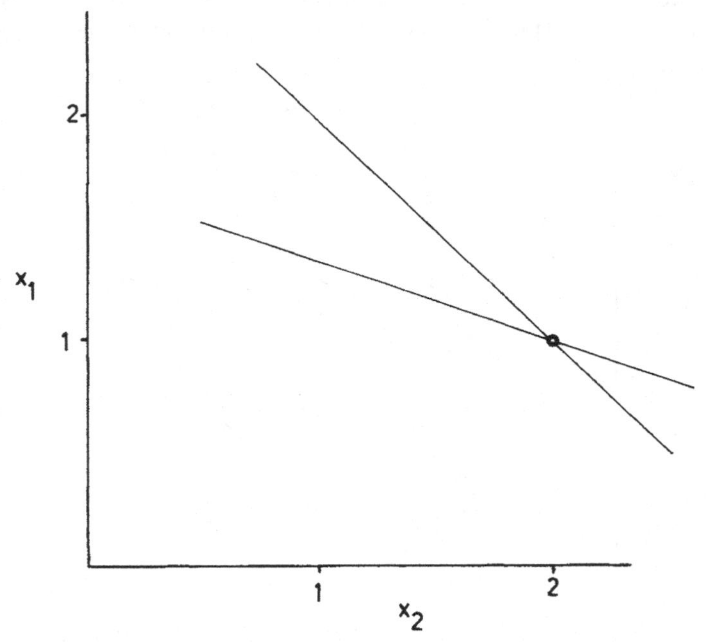


Fig. 9-6. Simultaneous equations intersecting at the point represented by the solution set.

### Exercise 9-13

Solve the simultaneous Eq. (9-47) by Gaussian substitution.

Solution 9-13. The Gaussian substitution method of solving simultaneous equations should be familiar from elementary algebra. First, express one of the unknowns in terms of the other

$$x_2 = 5 - 3x_1$$

then substitute into the second equation

$$x_1 + (5 - 3x_1) = 3$$

Solve for the other unknown from the second equation

$$x_1 + 5 - 3x_1 = 3$$
$$-2x_1 = -2$$
$$x_1 = 1$$

Now substitute that value back into the first equation

$$3(1) + x_2 = 5$$
$$x_2 = 5 - 3 = 2$$

which leads to the solution set (1, 2).

## Solution of Simultaneous Linear Equations by Cramer's Rule

In general, any number of simultaneous equations in any number of unknowns can be set up in the following notation

$$a_{11}x_1 + a_{12}x_2 + \cdots + a_{1N}x_N = b_1$$
$$a_{21}x_1 + a_{22}x_2 + \cdots + a_{2N}x_N = b_2$$
$$\vdots$$
$$a_{N1} + a_{N2}x_2 + \cdots + a_{NN}x_N = b_N \tag{9-48}$$

The constants $a_{11}$, $a_{12}$...are the *coefficients of the linear equations*, 3, 1, 1, and 1 in the Eq. set (9-47). The unknowns are $x_1$, $x_2$..., and the constants to which each equation is equal, $b_1$, $b_2$..., are called the *nonhomogenous* part of the set or the *nonhomogeneous vector*. In the simple Eq. set (9-47), the nonhomogeneous vector is 5, 3. Simultaneous equations may be set up in matrix form

$$\begin{bmatrix} a_{11} \; a_{12} \cdots a_{1N} \\ a_{21} \; a_{22} \cdots a_{2N} \\ \cdot \\ \cdot \\ \cdot \\ a_{N1} \; a_{N2} \cdots a_{NN} \end{bmatrix} \begin{bmatrix} x_1 \\ x_2 \\ \cdot \\ \cdot \\ \cdot \\ x_N \end{bmatrix} = \begin{bmatrix} b_1 \\ b_2 \\ \cdot \\ \cdot \\ \cdot \\ b_N \end{bmatrix} \tag{9-49}$$

where the first block of numbers containing the values of $a$ is called the *matrix of the coefficients*, the column containing $x_1$, $x_2$ ...$x_N$ is called the *solution vector*, and the column containing the $b$ values is, as we have seen, the nonhomogeneous vector. In general, vertical columns of numbers are called vectors and rectangular arrays are called *matrices* (singular, matrix). The use of the term rectangular "array" for a matrix is embedded in the literature of matrices; it has nothing to do with the order in which the *elements* of the matrix are arranged in rectangular form and does not imply that they are arrayed from smallest to largest.

## The Determinant of the Coefficients

Determinants are written in a notation that is similar to that for matrices, but the two are very different mathematical entities. As this point, it is useful to go into the differences between *scalars*, vectors, and matrices in a limited way. Scalars are numbers having magnitude only, that is, size. Vectors have both magnitude and direction. An example of a scalar is the speed of a molecule, which can have any value from zero to infinity, but is the same whether the molecule is travelling north, south, or some other direction. Velocity is a vectorial quantity and expresses both speed and direction, the vector 20 m $s^{-1}$ N (twenty meters per second North) being different from the vector 20 m $s^{-1}$ S (twenty meters per second South). The difference between vectors and scalars may be very important in some physical situations, for example, in map reading. Scalars are represented by collections of digits, e.g., 2 or 3.776..., vectors are represented as one-dimensional arrays, vertical arrays as written above, but equally well as one-dimensional horizontal arrays, $(x_1, x_2 \ldots x_N)$. Matrices are represented by two-dimensional rectangular arrays as above and cannot be reduced to any simpler form.

By contrast, *determinants* can be represented by two-dimensional arrays, but they are scalars and can always be reduced to a single number. Reduction of a determinant to a scalar is illustrated by the following example. The determinant, $D$,

$$D = \begin{vmatrix} 1 & 2 \\ 3 & 4 \end{vmatrix} \tag{9-50}$$

is written in rectangular form as one would write a matrix. The substitution of vertical lines for brackets in Eq. 9-50 serves to distinguish between the determinant, $D$, and the matrix, $A$

$$A = \begin{bmatrix} 1 & 2 \\ 3 & 4 \end{bmatrix} \tag{9-51}$$

Reduction of the determinant (9-50) is done by multiplying the elements on the principal diagonal and subtracting the product obtained by multiplying the remaining two elements. The *principal diagonal* of any matrix or determinant is the diagonal that begins at the upper left corner and stretches to the lower right corner.

$$D = \begin{vmatrix} 1 & 2 \\ 3 & 4 \end{vmatrix} = (1 \times 4) - (2 \times 3) = 4 - 6 = -2 \tag{9-52}$$

The determinant considered above is square and has two elements on an edge. It is, by analogy to matrix nomenclature, called a 2 × 2 determinant. Determinants arising in the solution of two equations in unknowns are 2 × 2 determinants. The general case,

$$D = \begin{vmatrix} a_{11} & a_{12} \\ a_{21} & a_{22} \end{vmatrix} \tag{9-53}$$

is expanded to

$$D = a_{11}a_{22} - a_{21}a_{12} \tag{9-54}$$

The next more complicated case is the 3 × 3 determinant, which is expanded by the method of *signed minors*. This can be done in several ways, but a simple one is to pick the top left element in the determinant, draw a line through it horizontally and another vertically

$$\begin{array}{c} \downarrow \\ \rightarrow \begin{vmatrix} a_{11} & a_{12} & a_{13} \\ a_{21} & a_{22} & a_{23} \\ a_{31} & a_{32} & a_{33} \end{vmatrix} \end{array} \tag{9-55}$$

Now, write the product of the element chosen times the determinant of all the elements not crossed out

$$a_{11} \begin{vmatrix} a_{22} & a_{23} \\ a_{32} & a_{33} \end{vmatrix} \tag{9-56}$$

This is repeated with the $a_{12}$ element and the $a_{13}$ element

$$\begin{array}{c} \phantom{a_{11}\ } \downarrow \phantom{\ a_{13}} \\ \rightarrow \begin{vmatrix} a_{11} & a_{12} & a_{13} \\ a_{21} & a_{22} & a_{23} \\ a_{31} & a_{32} & a_{33} \end{vmatrix} \end{array} \tag{9-57}$$

and

$$\begin{array}{c} \phantom{a_{11}\ a_{12}\ } \downarrow \\ \rightarrow \begin{vmatrix} a_{11} & a_{12} & a_{13} \\ a_{21} & a_{22} & a_{23} \\ a_{31} & a_{32} & a_{33} \end{vmatrix} \end{array} \tag{9-58}$$

leading to

$$a_{12} \begin{vmatrix} a_{21} & a_{23} \\ a_{31} & a_{33} \end{vmatrix} \tag{9-59}$$

and

$$a_{13} \begin{vmatrix} a_{21} & a_{22} \\ a_{31} & a_{32} \end{vmatrix} \tag{9-60}$$

Now, take the sum of all three terms obtained in this way and add them algebraically starting with a plus sign and alternating signs

$$a_{11} \begin{vmatrix} a_{22} & a_{23} \\ a_{32} & a_{33} \end{vmatrix} - a_{12} \begin{vmatrix} a_{21} & a_{23} \\ a_{31} & a_{33} \end{vmatrix} + a_{13} \begin{vmatrix} a_{21} & a_{22} \\ a_{31} & a_{32} \end{vmatrix} \tag{9-61}$$

We already know how to evaluate the three $2 \times 2$ determinants in the sum; hence we merely multiply them by the numbers $a_{11}$, $-\ a_{12}$, and $a_{13}$, add and obtain the $3 \times 3$ determinant we seek.

Though we shall not need anything larger than a $3 \times 3$ determinant in this book, it should be evident that the process shown has no limit. A $4 \times 4$ determinant can be broken down into a sum containing only $3 \times 3$ determinants and treated as above. A $5 \times 5$ determinant can be broken down to a sum containing only $4 \times 4$ determinants and so on.

The process may become very lengthy but, in principle, there is no determinant that can not be evaluated.

## Exercise 9-14

Evaluate the determinant

$$Det\ A = \begin{vmatrix} 3 & -1 & 0 \\ 5 & 2 & 4 \\ 1 & -3 & 1 \end{vmatrix}$$

Solution 9-14.

$$Det\ A = 3\begin{vmatrix} 2 & 4 \\ -3 & 1 \end{vmatrix} + \begin{vmatrix} 5 & 4 \\ 1 & 1 \end{vmatrix} + 0$$
$$= 3(2 + 12) + (5 - 4)$$
$$= 3(14) + 1 = 43$$

## Cramer's Rule

If we replace the $i$th column of the $N \times N$ coefficient matrix with the nonhomogeneous vector, we get $N$ new matrices that have the determinants $D_1, D_2, \ldots, D_i, \ldots D_N$. We shall denote the determinant of the matrix of the coefficients as $D$. Cramer's rule states that the solution set for an $N \times N$ set of simultaneous equations is $x_1 = D_1/D$, $x_z = D_2/D \ldots x_i = D_i/D \ldots x_N = D_N/D$.

### Exercise 9-15

Solve the simultaneous set

$$3x_1 + x_2 = 5$$

$$x_1 + x_2 = 3$$

by Cramer's rule.

Solution 9-15. Replacing column 1 of the coefficient matrix with the nonhomogeneous vector, we get

$$D_1 = \begin{vmatrix} 5 & 1 \\ 3 & 1 \end{vmatrix}. \quad 5 - 3 = 2$$

Replacing column 2,

$$D_2 = \begin{vmatrix} 3 & 5 \\ 1 & 3 \end{vmatrix} = 9 - 5 = 4$$

and

$$D = \begin{vmatrix} 3 & 1 \\ 1 & 1 \end{vmatrix} = 3 - 1 = 2$$

hence

$$x_1 = D_1/D = 2/2 = 1 \qquad x_2 = D_2/D = 4/2 = 2$$

which is the solution set. The treatment above presupposes that $D$ is not zero because division by zero is undefined.

Cramer's rule, though it may not be the best method of solving indefinitely large sets of simultaneous equations is certainly an *in principle* solution of the problem. There is no limit on the dimensions of the determinants that can be expanded and solved by the method of signed minors so long as they are square. Hence there is no limit on the number of equations that can be solved by Cramer's rule, provided that $D \neq 0$. The difficulty with Cramer's rule is that the expansion of determinants by the method of signed minors rapidly becomes more time consuming as the size of the determinant increases. Three by three determinants with integers as elements are easily expanded, $4 \times 4$ determinants are somewhat more difficult, but at the level of $5 \times 5$ or higher, the method becomes sufficiently tedious that the researcher becomes discouraged from using Cramer's rule, particularly since research data are almost never integral and may contain many significant figures that must be carried through the computations and into the final result. These practical considerations do not in any way vitiate the generality of the method. Computer methods exist for evaluation of determinants, but they are not widely used. Though we shall use Cramer's rule to solve the normal equations below, and in Chapter 9, matrix inversion methods are preferred for treatment of numerical data.

## Solution of the Normal Equations for Linear Curve Fitting by Cramer's Rule

We may now return to the normal equations, Eqs. (9-42) and (9-43) which can be written

$$N(b) + \sum x_i(m) = \sum y_i \tag{9-62}$$

and

$$\sum x_i(b) + \sum x_i^2(m) = \sum x_i y_i \tag{9-63}$$

so as to emphasize that the solution set is $(b, m)$, the nonhomogeneous vector is $\{\Sigma y_i, \Sigma x_i y_i\}$ and the matrix of the coefficients is

$$\begin{bmatrix} N & \sum x_i \\ \sum x_i & \sum x_i^2 \end{bmatrix} \tag{9-64}$$

with a determinant of coefficients of the same form. Applying Cramer's rule, we wish to find

$$b = D_b/D \tag{9-65}$$

and

$$m = D_m/D \tag{9-66}$$

corresponding to the general set of Cramer's rule equations, $x_1 = D_1/D$ and $x_2 = D_2/D$ that we developed in the previous section. Now, for this set,

$$D_b = \begin{vmatrix} \sum y_i & \sum x_i \\ \sum x_i y_i & \sum x_i^2 \end{vmatrix} \tag{9-67}$$

where we have substituted the nonhomogeneous vector for column 1 in $D$,

$$D_m = \begin{vmatrix} N & \sum y_i \\ \sum x_i & \sum x_i y_i \end{vmatrix} \tag{9-68}$$

where we have substituted the nonhomogeneous vector for column 2 of $D$ and

$$D = \begin{vmatrix} N & \sum x_i \\ \sum x_i & \sum x_i^2 \end{vmatrix} \tag{9-69}$$

which corresponds to the matrix 9-64. Expanding,

$$D_b = \sum y_i \sum x_i^2 - \sum x_i \sum x_i y_i$$
$$D_m = N \sum y_i x_i - \sum y_i \sum x_i$$
$$D = N \sum x_i^2 - \left(\sum x_i\right)^2 \tag{9-70}$$

Solving Eqs. (9-65) and (9-66) for $b$ and $m$, we have

$$b = D_b/D = \left[\sum y_i \sum x_i^2 - \sum x_i \sum x_i y_i\right] \Big/ \left[N \sum x_i^2 - \left(\sum x_i\right)^2\right] \tag{9-71}$$

(i.e., Eq. 9-45)

and

$$m = D_m/D = \left[N \sum x_i y_i - \sum y_i \sum x_i\right] \Big/ \left[N \sum x_i^2 - \left(\sum x_i\right)^2\right] \tag{9-72}$$

(i.e., Eq. 9-44)

which are the solutions of the normal equations we sought. Numerous applications of these rules and equations will be seen in the next chapter.

## Glossary

*Absorbance*. The amount of light taken up by a substance, usually in solution. Strictly, the logarithm of the ratio of radiative power impinging on a sample to that emerging from it, $A = \log p_o/p$.

*Absorbancy*. Ratio of the absorbance to the concentration, $a = A/c$ in a 1 cm cell.

*Beer's Law.* Basic law of spectroscopy relating absorbance $A$, light path $b$, absorbancy $a$ and $c$ concentration; $A = abc$.
*Coefficient Matrix.* Rectangular array consisting of the premultiplying coefficients in a set of simultaneous equations.
*Correlation Coefficient.* Statistic, $r$, designed to test correlation of variables, having values 0 to $\pm 1$ where zero is no correlation, $+1$ is perfect correlation, and $-1$ is perfect negative correlation. Sometimes called "Pearson's $r$."
*Cramer's Rule.* Method of solving simultaneous equations by evaluating determinants.
*Determinant.* Scalar quantity corresponding to a matrix.
*Linear Function.* Function $y = f(x)$ for which all values fall on a straight line, i.e., function with a constant slope.
*Matrix.* Rectangular array of elements; in this treatment, the elements are numbers or sums.
*Minimization Parameter.* Empirical parameter adjusted to minimize some function, in this case, the sum of squares of the deviations.
*Minor.* Rectangular array obtained by eliminating one row and one column of a larger rectangular array.
*Nonhomogeneous Vector.* Ordered set of the constant terms in a set of simultaneous equations.
*Origin.* In cartesian coordinates, the point at which $x = y = 0$.
*Partial Differentiation.* Differentiation of a function of two or more variables with respect to one of the variables, holding all others constant (see Chapter 2).
*Principal Diagonal.* That diagonal of a rectangular array starting at the top left and ending at the bottom right.
*Principle of Least Squares.* Principle that the best curve representing a data set minimizes the sum of squares of the deviation of each point from the curve.
*Residual.* The difference between $y_i'$ as calculated from a least squares function through $n$ points and $y_i$, the actual $i$th experimental value.
*Scalar.* Number having magnitude, but not direction, e.g., temperature.
*Simultaneous Equations.* Two or more linear equations under the constraint that they be simultaneously true. The number of solutions for which this constraint is satisfied may be one, none or infinitely many.
*Slope.* Change in $y$, $\Delta y$ divided by the change in $x$, $\Delta x$ where the changes are small, strictly speaking, infinitesimally small.
*Solution Set.* All the solutions of a set of simultaneous equations.
*Vector.* Number having magnitude and direction, e.g., force.

## Problems

*1.* Using the method of Exercise 9-1, determine which of the three values of $\mu$, 17.6 17.9, 18.1 best approximates $\bar{x}$ for the data set 17.4, 18.1, 17.7, 19.3, 16.2, and 18.6. What is the true value for $\bar{x}$.

2. Evaluate the determinant

$$\begin{vmatrix} 1 & 5 & 9 \\ 3 & 4 & 1 \\ 2 & 4 & 6 \end{vmatrix}$$

*3*. Evaluate the determinant

$$\begin{vmatrix} a & b & c \\ b & a & c \\ b & c & a \end{vmatrix}$$

*4*. Solve the simultaneous equations

$$2x + y = 8$$

$$x + 5y = 11$$

by Cramer's rule.

*5*. Solve the equations

$$x_1 + x_2 + x_3 = 5$$

$$x_1 + 3x_2 + x_3 = -1$$

$$2x_1 + 2x_2 + x_3 = 3$$

by Cramer's rule.

*6*. Find the least squares trend line for the following data on annual sales versus year in production for a certain product.

| Year in production | 1 | 2 | 3 | 4 | 5 |
|---|---|---|---|---|---|
| Sales in millions | 3 | 5 | 7 | 9 | 11 |

*7*. What is Pearson's *r* coefficient (the correlation coefficient) for the data in problem 6?

*8*. An organic dye has the following absorbancies at the corresponding concentrations

| *A* | 0.120 | 0.223 | 0.372 | 0.455 | 0.601 |
|---|---|---|---|---|---|
| Conc(*M*) | $1 \times 10^{-5}$ | $2 \times 10^{-5}$ | $3 \times 10^{-5}$ | $4 \times 10^{-5}$ | $5 \times 10^{-5}$ |

What is the equation of the least squares straight line passing through the origin for these points? What is the standard deviation of the residuals in *A*? If a solution of the same dyestuff has $A = 0.307$, as measured in the same optical cell, what is its concentration calculated from the least squares line?

*9*. Two different methods were used to determine parts per million (ppm) of calcium in blood serum. The results were

| Method 1 (ppm) | 2.4 | 4.8 | 6.2 | 8.5 | 10.0 |
|---|---|---|---|---|---|
| Method 2 (ppm) | 1.0 | 3.1 | 4.1 | 5.3 | 5.3 |

Are the results well correlated? Suppose that the true values are 2.0, 4.0, 6.0, 8.0, and 10.0 ppm, which appears to be the better method?

## Bibliography

P. M. Cohn, *Linear Equations*, Dover, New York, 1958.

J. T. Schwartz, *Introduction to Matrices and Vectors*, Dover, New York, 1972.

F. Ayres, Jr., *Theory and Problems of Matrices*, Schaum, New York, 1962.

S. Chatterjee and B. Price, *Regression Analysis by Example*, Wiley, New York, 1977.

# Chapter 10

# Fitting Nonlinear Curves

Although linear functions are very commonly encountered in research, the reader will find it easy to think of situations in which experimental data do not conform to a linear function. Hyperbolic behavior is familiar from Boyle's law, exponential population growth is familiar to students of ecology, and exponential radioactive decay is familiar to the radiobiologist.

Mathematical methods can be developed to treat nonlinear functions without previous manipulation, but they are generally fairly complicated. A convenient and useful strategy is to convert a nonlinear function into a linear one and apply the methods we already have to obtain the empirical parameters characterizing the function. The method will be introduced by treating Boyle's law, a hyperbolic function, in such a way as to obtain a linear function through the origin with $k$ as the slope. The new function will be treated by the method described in the last chapter to obtain the best value for $k$. Knowing $k$, we may calculate a value of $V$ for each of a series of arbitrarily selected values of $P$ (or vice versa), and so obtain the "best" hyperbola through the experimental points. The same strategy will then be applied to a number of other functions commonly occuring in the life sciences. Finally, a system will be developed for fitting a curve to a set of experimental points that are nearly on a straight line, but show a regular and increasing trend away from linearity. Such curves occur frequently in experimental work and are often well approximated by a parabola. We shall employ the Cramer's rule method of Chapter 9 to obtain an expression that permits calculation of the three experimental parameters resulting from the parabolic curve-fitting method. The equations produced by this treatment are by far the most cumbersome to be encountered in this book and computer treatment is a virtual necessity.

## Boyle's Law

The usual expression for Boyle's law, $PV = k$ where $P$ is the pressure of a gas, $V$ is its volume, and $k$ is a constant, leads to one branch of a hyperbola

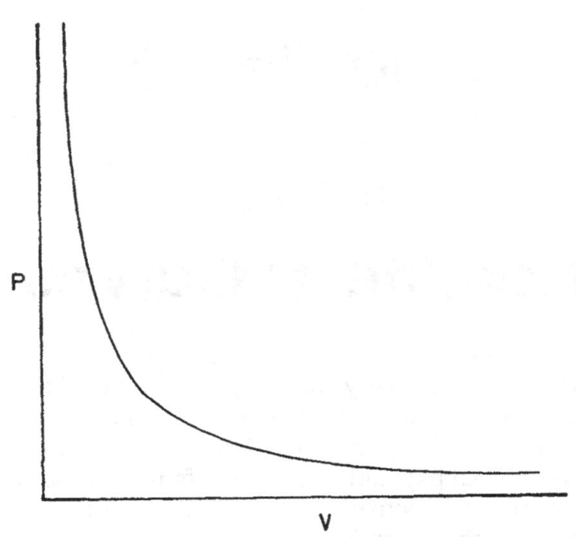


Fig. 10-1. Ideal Boyle's law curve.

in the upper right-hand quadrant of a flat surface representing all possible values of $P$ and $V$. Since negative values of $P$ and $V$ are physically meaningless, the only quadrant of physical interest is the one with both $P$ and $V$ positive.

Experimental measurements do not, of course, provide an ideal Boyle's law curve, so the usual question arises about how to draw the best curve through a set of points exhibiting experimental scatter. One way is to transform the hyperbolic function, $PV = k$, into a linear function by a simple substitution of variables. Algebraic rearrangement and grouping of variables leads to the following set of equivalent equations: $PV = k$, $P = k/V$, and $P = k\,(1/V)$. If we make the substitution of variables $Y = 1/V$, we have

$$P = kY \qquad (10\text{-}1)$$

which is the equation of a linear function passing through the origin. Least squares treatment of $P$ as a function of $Y$ leads to a best value of the slope, $k$.

Suppose that a cylinder is filled with gas, adjusted to eight different values of $V$ by means of a piston and that the pressure is read from a suitable gage. A value of $P$ exists for every value of $V$ and *vice versa*. Corresponding values of $P$, $V$, and $Y$ are given in the first three columns of Table 10-1.

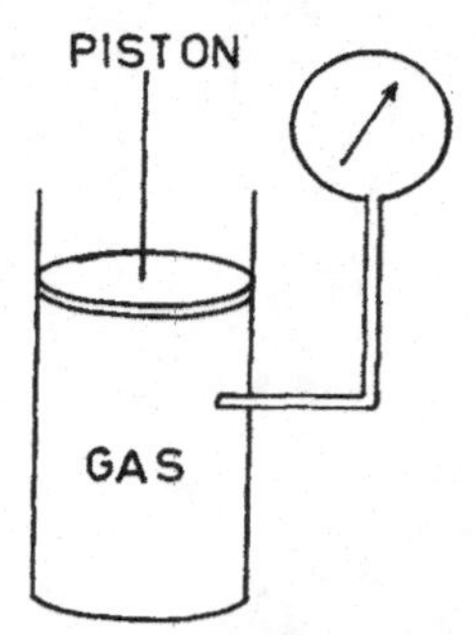


Fig. 10-2. Schematic diagram of an apparatus for measuring $P$ as a function of $V$.

Table 10-1
Corresponding Values of Pressure, Volume, and Reciprocal Volume Recorded by the Apparatus Shown in Fig. 10-2

| $P$ atmospheres | $V$, liters | $Y$, liters$^{-1}$ | $P$ corrected, atmospheres |
|---|---|---|---|
| 0.55 | 4.00 | 0.25 | .632 |
| 0.60 | 5.00 | 0.20 | .506 |
| 0.95 | 3.00 | 0.33 | .843 |
| 1.35 | 2.00 | 0.50 | 1.26 |
| 1.80 | 1.50 | 0.67 | 1.69 |
| 2.80 | 1.00 | 1.00 | 2.53 |
| 4.00 | 0.50 | 2.00 | 5.06 |
| 5.00 | 0.65 | 1.54 | 3.90 |

The values in the first two columns are plotted in Fig. 10-3. They describe a hyperbolic function, but scatter is present. The value of $P$ as a function of $Y$, Fig. 10-4, shows fairly close adherence to a straight line, once again, with scatter.

Computer treatment of $P = f(Y)$, the data presented in columns 1 and 3 in Table 10-1 yields the empirical "best" linear equation

$$P = 2.53\ Y \tag{10-2}$$

where the constant 2.53 is the slope of the linear function in Fig. 10-4. Taking this value of $k$, it is possible to determine what the value of $P$ "should have been" for each value of $V$ in column 2 of Table 10-1. These values calculated from the least squares value of $k$ are listed in column 4 of Table 10-1 as corrected values of $P$. They represent values of $P$ that have

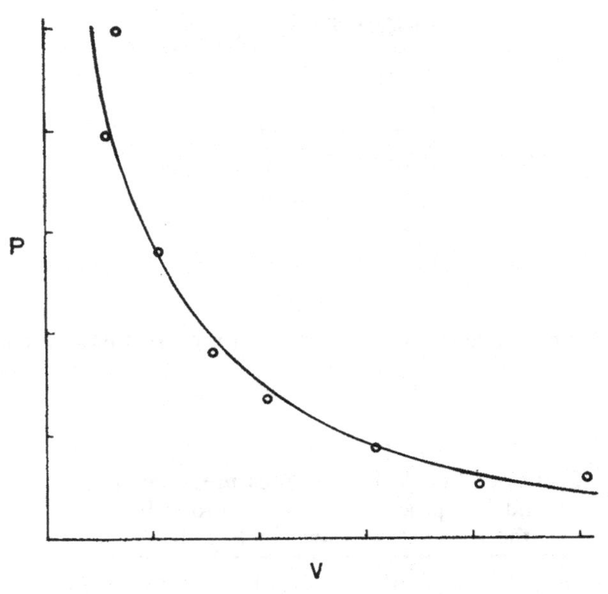


Fig. 10-3. The values of $P = f(V)$ for the data presented in Table 10-1.

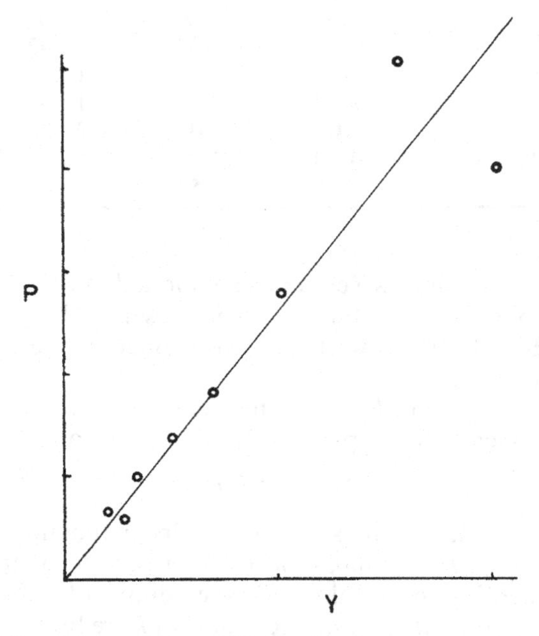


Fig. 10-4. The values of $P = f(Y)$ for the data presented in Table 10-1.

been "smoothed" so as to minimize the effects of experimental scatter by the least squares technique. The smooth curve drawn through these points is the hyperbola which best represents the experimental data.

## Exponential Decay

Many experimental curves decrease according to the equation

$$y = e^{-x} \tag{10-3}$$

This is a natural consequence of the mathematics of an event that is completely random in nature. The disintegration of a nucleus of the radioactive isotope of phosphorus, $^{32}P$, is a case in point that has many applications in the life sciences. When a nucleus of $^{32}P$ disintegrates, it gives off a β particle that can be detected by means of a Geiger counter

$$^{32}P \rightarrow {}^{32}S + \beta$$

When a collection of $^{32}P$ nuclei are investigated, the number of β particles emanating from the sample per unit time (second or minute) can be counted. This is the disintegration rate and is fairly reproducible from one sample to the next, provided that the same amount of $^{32}P$ is present in each. Lack of exact reproducibility from one sample to the next owes to the statistical nature of radioactive decay. If we concentrate our attention on one nucleus of $^{32}P$, we have no deterministic way of telling when it will disintegrate to produce $^{32}S$ and β. We can, however, predict the number of disintegrations among a large number of $^{32}P$ nuclei within reasonable confidence limits on statistical grounds. Radioactive disintegration is governed by the Poisson distribution and was covered in part in Chapter 6, which should be reviewed at this point. For present purposes, it is sufficient to note that the number of random disintegrations in a sample containing $10^{20}$ atoms and measured over a long time period will be about twice as great as the number of disintegrations of a sample containing $0.5 \times 10^{20}$ atoms, about four times as great as the number for a sample containing $0.25 \times 10^{20}$ atoms, and so on. In short, the predicted number of disintegrations is directly proportional to the amount of radioactive material present. Since each disintegration reduces the number of radioactive nuclei by one, we may write the disintegration rate as $-dN/dt$. The differential form signifies the rate of change of the number of nuclei of radioactive material, $N$, with respect to time, $t$, and the minus sign shows that $N$ is *decreasing* with time. In light of the proportionality between the rate of radioactive decay just mentioned and the amount of sample (number of radioactive nuclei) present, we may write

$$dN\big/dt \propto N \tag{10-4}$$

A proportionality sign $\propto$ may always be replaced by an equal sign and a constant of proportionality, $k$, to give

$$-dN/dt = kN \tag{10-5}$$

which is Eq.(6-16).

The proportionality constant, $k$, is characteristic of each radioactive element, being large if disintegration is relatively frequent in a given statistical grouping and small if it is infrequent. Knowing only $k$, called the *decay constant*, one can identify the material undergoing disintegration. The decay constant is also a measure of the probability of disintegration of individual nuclei of a given kind, hence it is a statistical parameter of some importance. For these reasons, we wish to extract the best value of $k$ we can obtain from any set of experimental measurements. Measurement of $dN/dt$ suffers from difficulties of both a statistical and experimental nature, hence we would not be wise to place our reliance in one measurement of $N$ and the rate as means to obtain $k$ from Eq. (10-5).

The most accurate value of $k$ is obtained by treating Eq. (10-5) so as to convert it into a linear function with $k$ as the slope, then to compute the slope by the least squares treatment for a linear function not passing through the origin as described in the previous chapter. This is done by rearranging Eq. (10-5) to obtain

$$dN/N = -kdt \tag{10-6}$$

When we take the indefinite integral of both sides,

$$\int dN/N = -k\int dt \tag{10-7}$$

we get

$$\ln N = -kt + \text{const.} \tag{10-8}$$

because the integral of $dN/N$ is the natural logarithm of $N$, $\ln N$. The constant at the right of Eq. (10-8) is called the constant of integration and appears whenever an indefinite integral is taken. Looking at Eq. (10-8) in search of a linear relationship, we see that if $\ln N$ is taken as the dependent variable and $t$ is taken as the independent variable, $\ln N = f(t)$ should be a linear function with the slope $-k$ and an intercept equal to the constant of integration. We have found the linear relationship we seek.

To illustrate the way in which experimental measurements are treated by this method, let us reconsider the decay of $^{32}P$ to produce a nonradioactive isotope of sulfur, $^{32}S$. Suppose, at the beginning of a series of measurements, a sample of $^{32}P$ produces 30,000 $\beta$ particles or "counts" per minute. This number is proportional to the number of atoms present in the sample at the beginning of the experiment, $N_0$. By substituting $t = 0$ into Eq. (10-8), we have the constant of integration

$$\ln N_0 = \text{const.} \tag{10-9}$$

After exactly 4.00 days, the count rate is checked again using a Geiger counter and found to be 24,500 counts per minute (cpm). After 8 days, the counter registers 20,500 cpm and so on until the data in the first and second columns of Table 10-2 have been recorded.

Table 10-2
Counts Per Minute Recorded After Various Time Intervals for a Sample of $^{32}P$

| $t$, days | cpm observed | ln cpm |
|---|---|---|
| 0 | 30,000 | 10.309 |
| 4 | 24,500 | 10.106 |
| 8 | 20,500 | 9.928 |
| 12 | 16,800 | 9.729 |
| 16 | 13,600 | 9.518 |
| 20 | 11,200 | 9.324 |
| 24 | 9,500 | 9.159 |
| 28 | 7,500 | 8.923 |

The third column of Table 10-2 gives the natural logarithm of the experimentally measured cpm values. The plot of experimental counts per minute as a function of time in days elapsed after the initial measurement is shown in Fig. 10-5. It has experimental scatter but is clearly not a straight line. Nor can the experimental curve in Fig. 10-5 be a hyperbola as in the previous section because the experimental curve cuts the vertical axis at $dN/dt = 30{,}000$ cpm. This is not possible for a true hyperbola which ap-

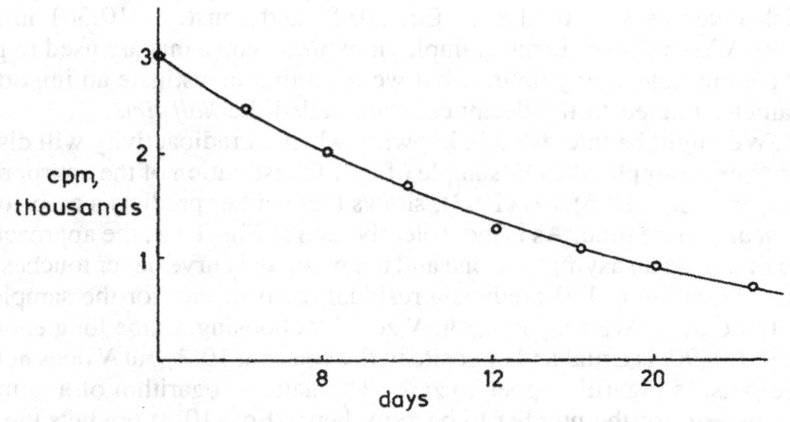


Fig. 10-5. Counts per minute as a function of time for a sample of radioactive phosphorus, $^{32}P$.

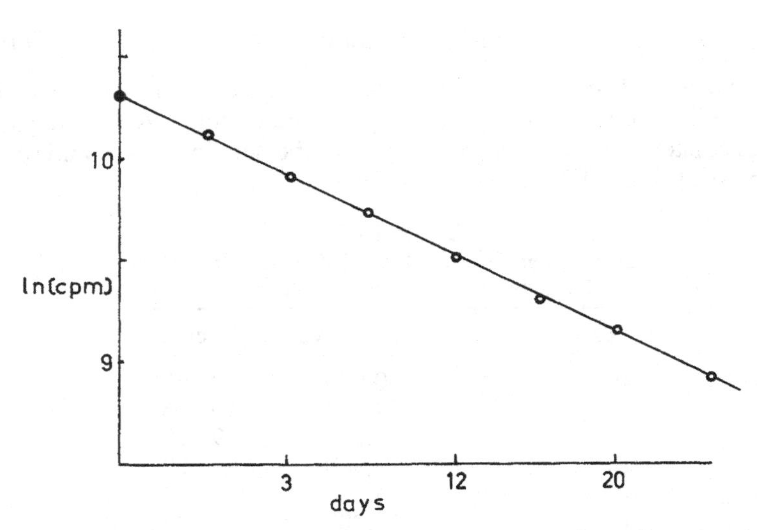


Fig. 10-6. Natural logarithm of cpm as a function of $t$ for radioactive decay of $^{32}P$.

proaches both the vertical and horizontal axes asymptotically (that is, the function approaches closer and closer to the axis but does not meet it).

The curve of ln $N$ versus $t$ is shown in Fig. 10-6. Although experimental scatter is still present, the function is evidently linear.

Computer treatment of the linear data in columns one and three of Table 10-2 leads to the equation

$$\ln(\text{cpm}) = -0.0490t + 10.301 \quad (10\text{-}10)$$

which gives us $k = 0.0490$ in Eq. (10-8) and const. $= 10.301$ in Eq. (10-9). We shall see, in an example, how these constants are used to predict a count rate at any time, $t$, but we must first investigate an important parameter related to the decay constant called the *half time*.

We might be interested in knowing when all radioactivity will disappear from a sample like our sample of $^{32}P$. Investigation of the appropriate equations, Eq. (10-6) and (10-8), shows that neither predicts a zero count rate at any finite time. As is most clearly seen in Fig. 10-5, the approach to zero of cpm is an asymptotic one and the predicted curve never touches the $t$ axis. Equation (10-8) predicts a residual radioactivity for the sample at any finite time. We may make ln $N$ zero by choosing a time long enough that $-0.0490t$ is equal and opposite to the constant 10.3, but $N$ does not go to zero as its logarithm goes to zero. The natural logarithm of a number must be $-\infty$ for the number to be zero; hence Eq. (10-8) predicts the absence of radioactive nuclei only at infinite time. This is true of all radioactive decompositions, hence the total decomposition time is of no value in characterizing or distinguishing radioactive elements.

A parameter that is useful in this respect is the half time or half life. If we start measuring time when the count rate is 30,000 cpm as we did in Fig. 10-5, and we measure the time necessary for the count rate to drop to 15,000 cpm, we have the half time, $t_{1/2}$. Inspection of Fig. 10-5 shows that the half time is about 14 days. If we start measuring time elapsed at 20,000 cpm and register $t_{1/2}$ at 10,000 cpm, we find it is about 14 days once again. Measuring the second half time after one has elapsed, i.e., from 15,000 to 7,500 cpm leads once again to $t_{1/2} = 14$ days. These three measurements of $t_{1/2}$ are shown in Fig. 10-7.

The half time is independent of the count rate and the time one starts counting. Independence of $t_{1/2}$ and the initial count rate is not a coincidence in Fig. 10-7, nor is it peculiar to $^{32}P$ decomposition. Rather, it results from the mathematical form of the first order decomposition as shown immediately below. Combining Eqs. (10-8) and (10-9), we have

$$\ln N = -kt + \ln N_0 \tag{10-11}$$

which can be rearranged to

$$\ln N - \ln N_0 = \ln N/N_0 = -kt \tag{10-12}$$

Since $\ln N/N_0 = -\ln N_0/N$, we may get rid of the minus sign and write

$$\ln N_0/N = kt \tag{10-13}$$

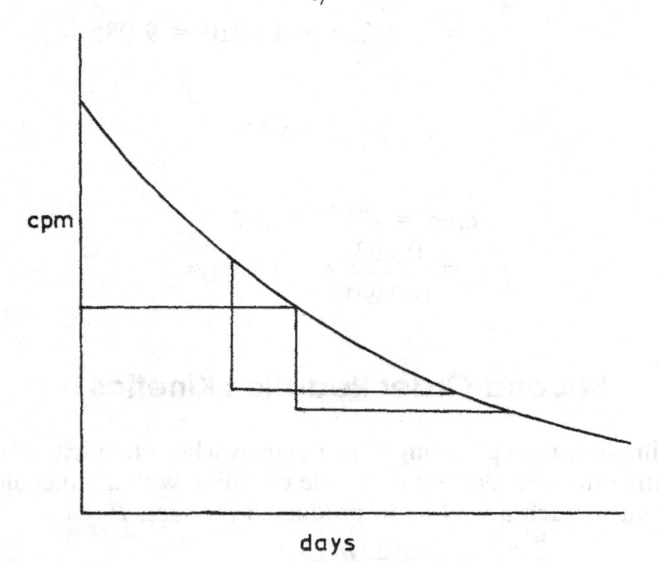


Fig. 10-7. Diagram showing the independence of $t_{1/2}$ and the count rate.

Now, the count rate is proportional to the number of radioactive nuclei present; hence if we denote the count rate at some arbitrarily selected zero time as $(\text{cpm})_0$, $N = \frac{1}{2}N_0$ when cpm $= \frac{1}{2}(\text{cpm})_0$. Substituting this condition, which holds only at $t_{1/2}$,

$$\ln\left(N_0 \big/ (1/2)N_0\right) = kt_{1/2} \tag{10-14}$$

or

$$\ln\left(1 \big/ (1/2)\right) = \ln 2 = kt_{1/2} \tag{10-15}$$

The tabulated value of ln 2 is 0.693; hence we have the simple relationship

$$kt_{1/2} = 0.693 \tag{10-16}$$

which is the same as Eq. (6-15).

### Exercise 10-1

Calculate the "best" estimate of the count rate at $t = 25.0$ days. Determine the best estimate obtainable from the data for the half life of $^{32}P$.

Solution 10-1. The best values of the count rate and $t_{1/2}$ obtainable from these data result from the computer-smoothed least squares parameters in Eq. (10-10). At $t = 25.0$ days,

$$\begin{aligned} \ln(\text{cpm}) &= -0.0490(25.0) + 10.310 \\ &= -\ 1.225 + 10.310 = 9.085 \end{aligned}$$

If

$$\ln(\text{cpm}) = 9.085$$

then

$$\text{cpm} = e^{9.085} = 8820$$

$$t_{1/2} = \frac{0.693}{0.0490} = 14.1 \text{ days}$$

## Second-Order Reaction Kinetics

One very simple way of picturing what happens when chemical substances react is to imagine one reactant molecule colliding with a molecule of the other reactant in such a way as to produce a product, $P$

$$A + B = P$$

Although most chemical reactions, when studied in detail, turn out to be more complicated than this, *bimolecular* collisions are quite important in all manner of reactions from combinations of the simplest atoms to the most complex reaction sequences in biological systems. When studied alone or in such a way that interferences can be screened out, reactions of this sort follow a *second-order* rate equation, Eq. (10-17). This is reasonable from statistical considerations. We suppose that the number of $A$–$B$ collisions will be simultaneously proportional to the amount of $A$ present and to the amount of $B$. If either reagent is in short supply, there will be few reactive collisions. Mathematically,

$$dP/dt \propto AB \tag{10-17}$$

where $P$ is the concentration of product formed at time $t$ and $A$ and $B$ are the concentrations of the two reactants at any instant, $t$. If we set $A = B$, we have the most favorable circumstances for $A$–$B$ collisions to take place because the probability of unfruitful $A$–$A$ and $B$–$B$ collisions is minimized. There is also the mathematical advantage that, upon specifying $A_0 = B_0$, that is, the concentrations of the reactants are specified equal to each other at time $t = 0$, the concentrations of $A$ and $B$ will be equal throughout the entire reaction. This is so because each molecule of $A$ that is used up to produce a molecule of $P$ is accompanied by the loss of a molecule of $B$, which has been consumed in the same transformation. Hence, Eq. (10-17) can be simplified to give

$$dP/dt = kA^2 \tag{10-18}$$

and, if we note that the rate of product formation is also the rate of reactant depletion, we may write

$$-dA/dt = kA^2 \tag{10-19}$$

Of the various equivalent forms of the second-order rate equation, this is the most convenient mathematically because it can be rearranged to give

$$-dA/A^2 = kdt \tag{10-20}$$

The right side of Eq. (10-20) is essentially the same as it was for the first order case and the integral of interest on the left is

$$\int dx/x^2 = \int x^{-2}dx = (-1)x^{-1} + \text{const.} \tag{10-21}$$

whence the integral form of the second order rate equation

$$-\int dA/A^2 = k \int dt \tag{10-22}$$

becomes

$$1/A = kt + \text{const.} \tag{10-23}$$

We would like to evaluate the constant of integration, which is carried out as it was before, by setting $A = A_0$ at $t = 0$. When this is done,

$$1/A_0 = \text{constant} \tag{10-24}$$

whence, substituting the value of the constant back into Eq. (10-23),

$$1/A = kt + 1/A_0 \tag{10-25}$$

or

$$1/A - 1/A_0 = kt \tag{10-26}$$

Once again, we have taken a fairly complicated mathematical expression and cast it in the form of a linear equation. Equation (10-23) shows that the rate constant is the slope of a plot of $1/A$ versus $t$. Since there is a constant in this linear function, the two-parameter linear curve-fitting program must be used, which yields the slope and intercept. The intercept, as we have shown, is $1/A_0$, the inverse of the initial concentration. If, instead of plotting $1/A$ as the dependent variable, we plot $1/A - 1/A_0$ (assuming the initial concentration to be known) the function has been converted into a form that passes through the origin so that the simpler one-parameter curve-fitting routine can be used.

An important example of a second-order reaction treated by these methods is the decomposition of solutions of ammonium cyanate by heat to form urea

$$NH_4^+ + CNO^- = NH_2CONH_2$$

The initial solution of ammonium and cyanate ions is made up by dissolving $NH_4CNO$ in water, thus assuring that the resulting concentration of $NH_4^+$ is exactly equal to that of $CNO^-$, a condition stipulated in our derivation of Eq. (10-25).

Suppose that a solution of $NH_4CNO$ of known concentration is brought to 70°C and samples are withdrawn after certain time intervals and analyzed for urea. Because each molecule of urea formed consumes exactly one $NH_4^+$ ion, the concentration of $NH_4^+$ ion, which we shall call $A$, must be its initial known concentration minus the urea which has been formed. Consequently, we are able to tabulate a series of concentrations of $NH_4^+$, each corresponding to a reaction time, $t$, as shown in column two of Table 10-3. Useful derivative functions of the concentration, $1/A$ and $1/A - 1/A_0$, are given in columns three and four of the same table. A plot of $A$ versus $t$ is given in Fig. (10-8).

Figure 10-9 shows $1/A$ plotted as a function of $t$, which gives a straight line and can be fitted by a linear, two-parameter least squares computer program. The result of such a curve-fitting procedure has the slope of the line as 0.0715 and its intercept as 2.01. From this we conclude that $k = 7.15 \times 10^{-2}$ and that $A_0 = 0.498$. The discrepancy between the computed intercept and the known value of $A_0$ results from the curve fitting technique. The computed intercept, along with the slope, gives the linear function that best fits all the experimental points, but does not necessarily pass through any one of them.

Table 10-3
Concentration of Ammonium Ion, $A$, Present at 70°C at Various Reaction Times

| $t$, min | $A$, mol. liter$^{-1}$ | $1/A$ | $1/A - 1/A_0$ |
|---|---|---|---|
| 0 | 0.500 | 2.00 | 0.0 |
| 10 | 0.385 | 2.60 | 0.60 |
| 20 | 0.282 | 3.55 | 1.55 |
| 30 | 0.233 | 4.29 | 2.29 |
| 40 | 0.207 | 4.83 | 2.83 |
| 50 | 0.178 | 5.62 | 3.62 |
| 60 | 0.160 | 6.25 | 4.25 |
| 70 | 0.140 | 7.14 | 5.14 |
| 80 | 0.136 | 7.35 | 5.35 |
| 90 | 0.117 | 8.55 | 6.55 |
| 100 | 0.108 | 9.26 | 7.26 |

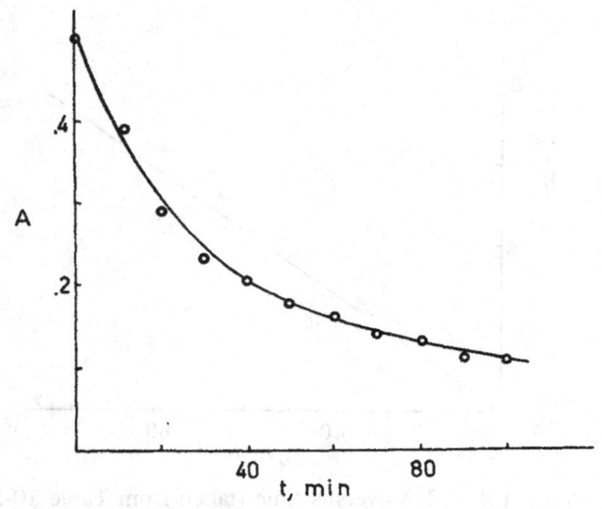


Fig. 10-8. Concentration of ammonium ion, $A$, versus time (see Table 10-3).

Figure 10-10 shows the alternate method of plotting $1/A - 1/A_0$ versus $t$ to obtain $k$. The slope is computed by a one-parameter method from Chapter 9. By the method used, the intercept must be zero. The value of $k$ obtained from Fig. 10-10 is slightly different from that obtained from Fig. 10-9 because of the difference between the one-parameter and the two-parameter least-squares fitting techniques.

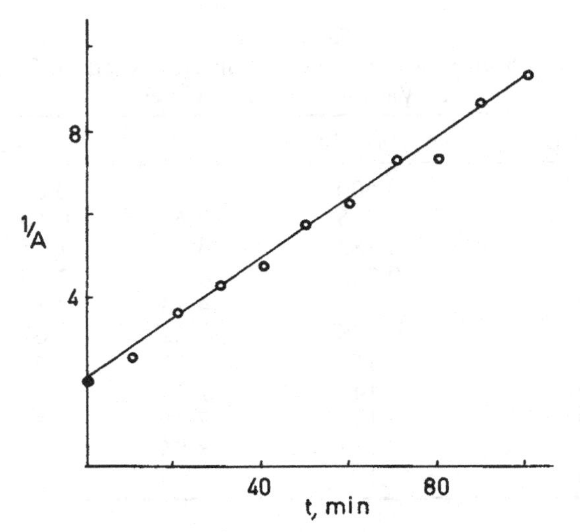


Fig. 10-9. Reciprocal of $A$ (Table 10-3) versus time.

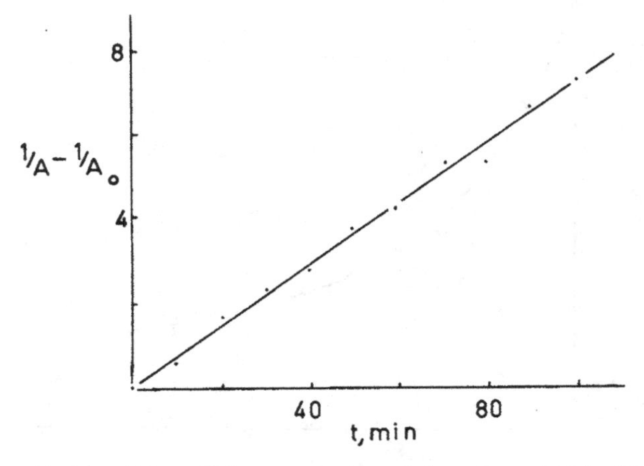


Fig. 10-10. $1/A - 1/A_0$ versus time (taken from Table 10-3).

### Exercise 10-2

Compute the slope of the function in Fig. 10-10 by a one-parameter method. Calculate the expected concentration of $A$ at 25.0 min reaction time.

## Enzyme-Catalyzed Reactions

Most reactions of real interest are a good deal more complicated than the simple cases of radioactive decay and bimolecular collision just described. One area of particular interest in the life sciences is that of *enzyme-catalyzed* reactions. In reactions of this kind, the rate of reaction of some *substrate* molecule is affected by the presence of an enzyme. It is customary to base one's calculations on the initial rate of reactions in studies of this kind to avoid interferences brought about by the reaction products. Idealized initial rate versus substrate curves are shown in Fig. 10-11.

The stepwise increase in maximum initial rate $R$ at three different enzyme concentrations, [E], 2[E], and 3[E], suggests a direct proportionality between rate, $R$, and enzyme concentration. The initial variation in rate with substrate concentration is linear, passes through the origin, and has slopes in the ratio 1:2:3 for the three enzyme concentrations. The rate equation for reactions of this kind is

$$R = k\,[E]\,[S]\big/(k' + [S]) \tag{10-27}$$

or

$$R = k\,[E]\big/(1 + k'\big/[S]) \tag{10-28}$$

where $k$ and $k'$ are empirical constants at this level of treatment. These

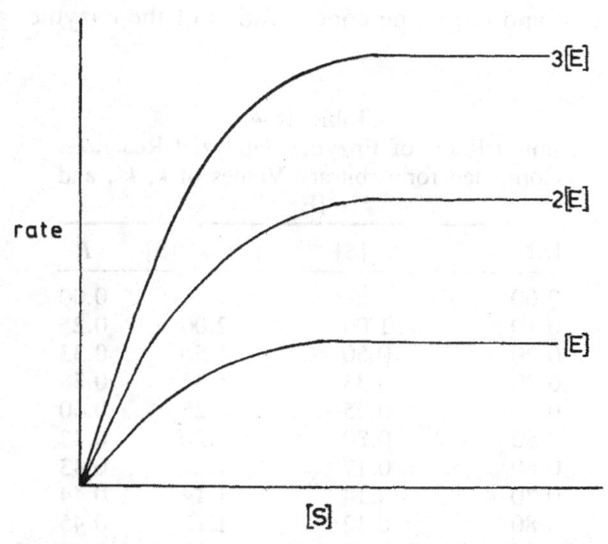


Fig. 10-11. Initial reaction rate as a function of substrate concentration [S] at three different enzyme concentrations [E].

constants have been interpreted on theoretical grounds (Barrow, 1974). Equation (10-27) is widely referred to as the *Michaelis-Menten equation.*

## Exercise 10-3

Taking the arbitrary values of $k = 0.5$ and $k' = 0.1$, compute the values for $R$ when [E] = 1.0 and [S]=0.0, 0.1, . . ., 1.0, i.e., [S] varies from 0 to 1 at intervals of 0.1. Plot the results and decide whether the general curve shape of Fig. 10-11 is followed.

Solution 10-3. Calculations from Eq. (10-28) lead to the results in Table 10-4.

The plot of $R$ versus [S] from Table 10-4 conforms to the general shape predicted for this type of enzyme-catalyzed reaction as shown in Fig. 10-12.

The practical problem actually encountered in enzyme kinetics studies is not constructing the experimental curve from a knowledge of the constants, but the reverse, that of obtaining the constants from a set of laboratory measurements. Suppose the initial rate were determined for a series of starch concentrations in an experiment on the rate of conversion of starch to sugar by the catalyst amylase. If the initial rates are found to be those in column three of Table 10-5, a plot resembling Fig. 10-12 is obtained. In this section, [S] should be taken to represent the concentration of starch substrate and [E] is the concentration of the enzyme, amylase.

Table 10-4
Initial Rates of Enzyme-Catalyzed Reactions Computed for Arbitrary Values of $k$, $k'$, and [E]

| [S], % | $k'$ [S] | 1 + $k'$ [S] | $R$ |
|---|---|---|---|
| 0.00 | | | 0.00 |
| 0.10 | 1.00 | 2.00 | 0.25 |
| 0.20 | 0.50 | 1.50 | 0.33 |
| 0.30 | 0.33 | 1.33 | 0.38 |
| 0.40 | 0.25 | 1.25 | 0.40 |
| 0.50 | 0.20 | 1.20 | 0.42 |
| 0.60 | 0.17 | 1.17 | 0.43 |
| 0.70 | 0.14 | 1.14 | 0.44 |
| 0.80 | 0.12 | 1.12 | 0.45 |
| 0.90 | 0.11 | 1.11 | 0.45 |
| 1.00 | 0.10 | 1.10 | 0.45 |

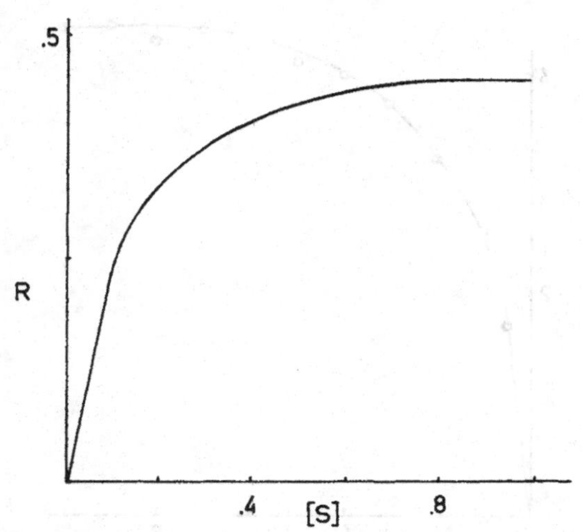


Fig. 10-12. Initial rate of enzyme-catalyzed reaction as a function of substrate concentration.

Table 10-5
Measured Initial Rates of the Amylase Conversion of Starch to Sugar

| [S] | 1/[S] | $R$ | $1/R$ |
|---|---|---|---|
| 0.050 | 20.0 | 0.17 | 5.88 |
| 0.100 | 10.0 | 0.26 | 3.85 |
| 0.200 | 5.00 | 0.32 | 3.08 |
| 0.300 | 3.33 | 0.37 | 2.67 |
| 0.400 | 2.50 | 0.40 | 2.50 |
| 0.500 | 2.00 | 0.41 | 2.44 |
| 0.600 | 1.67 | 0.42 | 2.35 |
| 0.700 | 1.43 | 0.44 | 2.27 |
| 0.800 | 1.25 | 0.43 | 2.33 |
| 0.900 | 1.11 | 0.44 | 2.25 |
| 1.000 | 1.00 | 0.44 | 2.25 |

Once again, we find it convenient to convert the empirical curve to a linear function. The reciprocal of Eq. (10-28) is

$$\frac{1}{R} = \frac{1 + k'/[\mathrm{S}]}{k[\mathrm{E}]} = \frac{1}{k[\mathrm{E}]} + \frac{k'}{k[\mathrm{E}]}\frac{1}{[\mathrm{S}]} \tag{10-29}$$

If the enzyme concentration, [E], is kept constant from one rate experiment to the next, $k$[E] is a constant, $(1/k)$[E] is a constant and $(k'/k)$[E] is

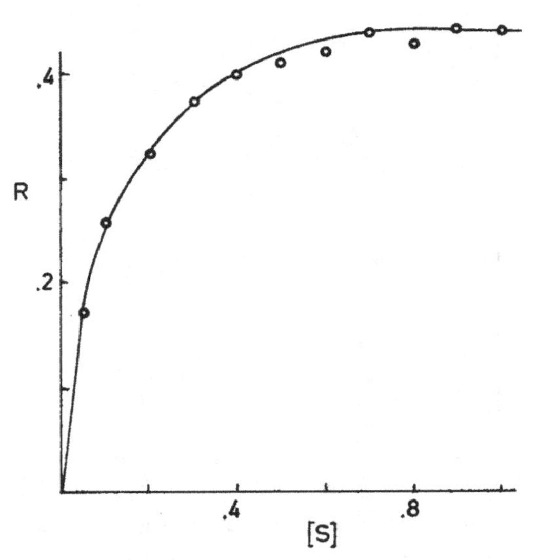


Fig. 10-13. Initial rates for amylase-catalyzed conversion of starch to sugar.

a constant. Denoting these latter two conglomerate constants, $K$ and $K'$ respectively, and rearranging Eq. (10-29) slightly, we have

$$1/R = K'(1/[S]) + K \tag{10-30}$$

Clearly, a plot of $1/R$ versus $1/[S]$ yields a straight line if the rate obeys Eq. 10-28. The linear function resulting from substitution of the data of Table 10-5 into Eq. (10-30) yields Fig. 10-14 called a Lineweaver-Burke plot.

Two immediate qualitative observations should be made on Fig. 10-14. First, although the function appears, at first glance, to pass through the origin, it does not. The $1/R$ axis extends over the range 2.0–6.0. Expansion of one or both axes is a common pictorial technique used to present graphs in the clearest way while avoiding wasted space. Second, mathematical treatment of the data has caused the data points to bunch up at one end of the curve. This is common in plotting techniques involving the reciprocal of one or both of the variables and we have seen it before in the analysis of Boyle's law. Treatment of data in this way places great weight on the one or two measurements at large values of $1/[S]$, i.e., small values of [S]. If there is a significant error in either of the rightmost data points in Fig. 10-14, the slope of the function will be affected. An error in one of the leftmost data points would have little effect on the slope. Looking back at Table 10-5, we see that the sensitive data points are those at lowest substrate concentration and the insensitive ones are at high substrate concen-

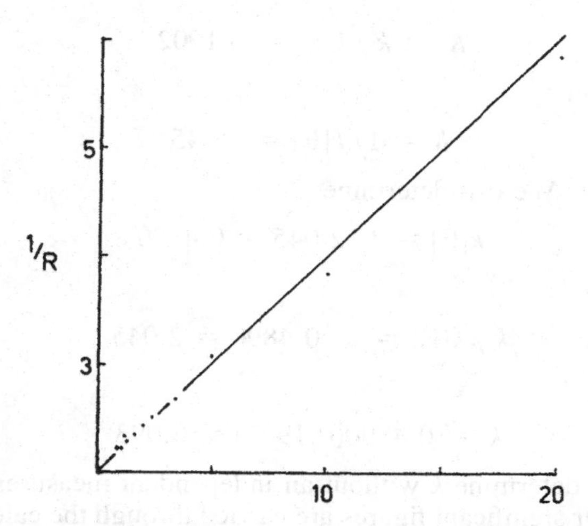


Fig. 10-14. $1/R$ versus 1/[S] for the amylase-catalyzed conversion of starch to sugar.

tration. This analysis should have an effect on our planning of experiments; in Lineweaver-Burke catalysis studies, experiments should be designed so as to obtain numerous and accurate data of the initial rate at low substrate concentrations. Weighting factors are often used to correct distortions of this kind.

## Exercise 10-4

Calculate approximate values for the slope and intercept of the linear function in Fig. 10-14 using only pencil and paper (no computers or calculators).

**Solution 10-4.** Taking the values of 1/[S] of 0 and 20, the corresponding values of $1/R$ are approximately 2.0 and 5.85. From elementary algebra, we calculate the slope of the function as

$$m = (5.85 - 2.00)/(20.0 - 0.0) = 3.85/20.0 = 0.192$$

The intercept is, by inspection, approximately 2.0.

Computer analysis of the data for $1/R$ and 1/[S] given in Table 10-5 yields the following analytical equation for the linear function shown in Fig. 10-14 in the form of Eq. (10-30)

$$1/R = 0.1902\ (1/[S]) + 2.045 \tag{10-31}$$

From this we have

$$K' = k'/k[\mathrm{E}] = 0.1902 \qquad (10\text{-}32)$$

as the slope and

$$K = 1/k[\mathrm{E}] = 2.045 \qquad (10\text{-}33)$$

as the intercept. We can determine

$$k[\mathrm{E}] = 1/2.045 = 0.4890 \qquad (10\text{-}34)$$

which leads to

$$k'/k[\mathrm{E}] = k'/0.4890 = 2.045 \qquad (10\text{-}35)$$

and

$$k' = 0.4890(0.1902) = 0.093 \qquad (10\text{-}36)$$

but we cannot determine $k$ without an independent measurement of [E]. Notice that four significant figures are carried through the calculation, but that there are only two given for $k'$. This is consistent with the precision of $R$ in Table 10-5.

## Unlimited Population Growth

The simplest theory of unlimited population growth depends on the postulate that reproduction and death are random phenomena with rates entirely dependent on the size of the population $N$. For growth of the population, the birth rate, $B$, must exceed the death rate, $D$. To simplify further, we shall take both $B$ and $D$ to be constant. Their difference,

$$\mathrm{U} = B - D \qquad (10\text{-}37)$$

is also constant and is called the *intrinsic rate of increase*. The rate of increase of the population owing to birth is

$$(dN/dt)_B = BN \qquad (10\text{-}38)$$

because of the random nature we have assumed for birth. Similarly, the rate of decrease in the population owing to death is

$$(-dN/dt)_D = DN \qquad (10\text{-}39)$$

or

$$(dN/dt)_D = -DN \qquad (10\text{-}40)$$

The rate of change of the entire population is the sum of the rate of increase owing to birth plus the rate of decrease owing to death

$$dN/dt = (dN/dt)_B + (dN/dt)_D = BN - DN$$
$$= (B - D)N = UN \qquad (10\text{-}41)$$

In the case of populations in general, the intrinsic rate of increase, $U$, may be positive, negative, or zero, but in the case that concerns us here, that of unlimited population growth, it is positive. Rearranging Eq. (10-41), we have

$$dN/N = Udt \tag{10-42}$$

which is solved in exactly the same way that the differential equation for exponential decay was solved to give

$$\ln N = Ut + \text{const.} \tag{10-43}$$

where the constant of integration can be evaluated by counting the population at a time designated as $t_0 = 0$. Now,

$$\ln N = Ut + \ln N_0 \tag{10-44}$$

or

$$\ln N/N_0 = Ut \tag{10-45}$$

Equation 10-45 is useful for purposes of computation, but it is not the form in which population growth curves are usually expressed. Perhaps we remember from algebra that a logarithm is an exponent, thus

$$10^{\log 5} = 5 \tag{10-46}$$

where we say that log 5 has the *base* 10. The natural logarithm, which appears whenever we integrate quotients like $dN/N$, is denoted ln and has the base 2.718. The number 2.718 appears so frequently in applied calculus that it is given a symbol, *e*. By analogy to Eq. (10-46),

$$e^{\ln 5} = \exp \ln 5 = 5 \tag{10-47}$$

In general, if $x = \ln y$,

$$e^x = e^{\ln y} = y \tag{10-48}$$

Applying this transformation to Eq. (10-45), if $Ut = \ln N/N_0$,

$$e^{Ut} = e^{\ln N/N_0} = N/N_0 \tag{10-49}$$

or

$$N = N_0 e^{Ut} \tag{10-50}$$

which is the form in which the analytical expression of exponential growth is usually written.

If we measure one arbitrary unit of time, $t$, Eq. (10-50) can be rearranged to give

$$N/N_0 = e^U \tag{10-51}$$

i.e., the ratio of population one unit of time after an arbitrarily chosen $t_0$ is always the same for a population undergoing unlimited growth. We may select our unit of time according to the clock, one minute, hour, day, etc., but it is more interesting to select it according to the population we are studying so as to make $N/N_0$ come out to a preselected ratio. We can select a time interval to make the ratio $N/N_0$ come out to any value we want from one ($t = 0$) to infinity ($t = \infty$), but the most common ratio is 2. When

$$2 = e^{Ut} \tag{10-52}$$

is solved for $t$, we have the time in which $N$ becomes twice $N_0$. This is called the *doubling time*.

Notice the analogous treatment of unlimited growth and radioactive decay. The doubling time of the former is analogous to the half time of the latter. When the natural logarithm of both sides of Eq. (10-52) is taken,

$$\ln 2 = 0.693 = Ut_2 \tag{10-53}$$

where $t_2$ indicates the doubling time. An accurate knowledge of either $U$ or the doubling time permits easy calculation of the other. If the exponential growth curve is perfectly smooth and obeys Eq. (10-53), it is , in principle, possible to determine $t_2$ by measuring the time necessary for the original population to double. In practice, this is not the best way to obtain either $t_2$ or $U$ because, with experimental scatter to contend with, it is not wise to base our empirical parameters on one experimental point. We must use the linear Eq. (10-43) or (10-44) as a basis for computer least-squares treatment of all the experimental data and so obtain the best value of $U$ and the doubling time.

## Exercise 10-5

A certain weevil is placed in an unlimited wheat supply at controlled temperature and humidity. There are 100 in the original colony. After one week there are 210, two weeks, 470, three weeks, 980, four weeks, 2060 and after five weeks there are 4480 in the colony. Calculate the intrinsic rate of increase, $U$, and the doubling time for this colony.

Solution 10-5. Computer least squares treatment of the equation

$$\ln N/N_0 = Ut$$

yields a slope of $U = 0.76$ weeks $^{-1}$. The doubling time is

$$0.693 = Ut_2$$

$$t_2 = 0.91 \text{ weeks}$$

## Limited Population Growth

No real population enjoys unlimited growth. Predators, disease, food, space, or other limitations eventually interfere. One mathematical model for handling population change (growth, decline, or stasis) is the expression of rate as a power series

$$dN/dt = a + bN + cN^2 + dN^3 + \ldots \quad (10\text{-}54)$$

which extends to an infinite number of terms, each preceded by a constant, $a, b, c, \ldots$ that may be positive, negative, or zero. The equation for population change is not soluble as it stands, but we can simplify it in two ways. First, terms on the right of Eq. (10-54) become less important as the exponent of $N$ increases; hence, if we cut off the infinite series at some finite number of terms, we have an equation that is approximately true. Second, we note that for a zero population, there can be no change. It must remain at $N = 0$ for it has no females to bear and no members of either sex to die. Substituting into Eq. (10-54)

$$dN/dt = a = 0 \quad (10\text{-}55)$$

hence $a$, being identically zero, may be dropped, leaving

$$dN/dt = bN + CN^2 + dN^3 + \ldots \quad (10\text{-}56)$$

If we drop all terms on the right of Eq. (10-56) except the first,

$$dN/dt = bN \quad (10\text{-}57)$$

we have the crudest or first approximation to population change afforded by this model. It is the case of unlimited population growth already discussed, where $b$ is the intrinsic rate of increase symbolized by $U$ in the previous section.

The second, and presumably more accurate, approximation to population growth is obtained by retaining two terms on the right of Eq. (10-56) rather than just one. The differential equation doesn't look much more difficult than the one we solved in the last section

$$dN/dt = bN + cN^2 \quad (10\text{-}58)$$

but it is. Solution is by a method known as series expansion, a technique beyond the scope of this book (Lotka, 1956; Cohen, 1933). The solution obtained in this way is

$$N = -(b/c)/(1 + e^{-bt}) \quad (10\text{-}59)$$

where $t$ is measured from the midpoint of the curve.

Table 10-6
Areas of a Bacterial Colony as a Function of Time

| Age of colony, days | Area, $cm^2$ | $(-B/C)/A$ | $\ln[(-B/C)/A] - 1$ |
|---|---|---|---|
| 0.0 | 0.24 | 208 | 5.33 |
| 1.0 | 2.78 | 18.0 | 2.83 |
| 2.0 | 13.53 | 3.70 | 0.99 |
| 3.0 | 36.30 | 1.38 | −0.97 |
| 4.0 | 47.50 | 1.05 | −2.94 |
| 5.0 | 49.40 | 1.01 | −4.41 |

Thornton, in 1922, observed the areas in column 2, Table 10-6, for a growing bacterial colony measured at intervals of one day up to 5.0 days.

Plotting area versus age of the colony in days leads to the sigmoid curve shown in Fig. 10-15. We wish, as usual, to convert this function to a linear one so as to compute the values of the constants.

Since colony area, $A$, is directly proportional to the number of bacteria in the colony, we may write the analog to Eq. (10-58) as

$$dA/dt = BA + CA^2 \tag{10-60}$$

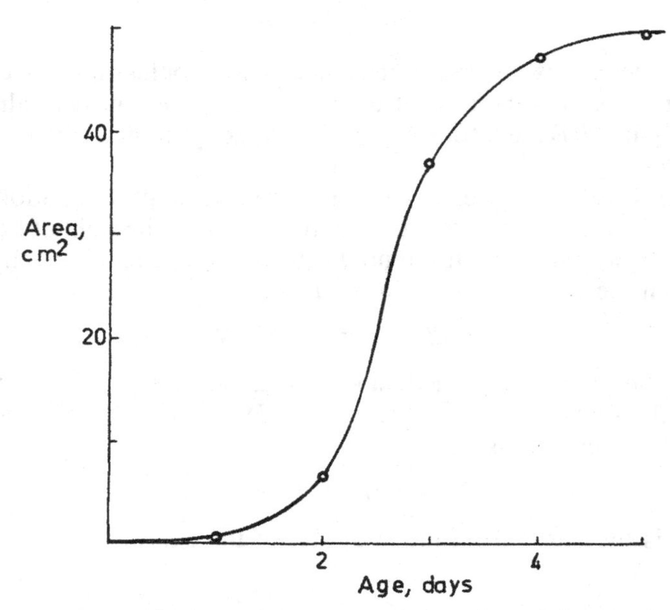


Fig. 10-15. Area of a bacterial colony as a function of time in days.

with the solution

$$A = (-B/C)\Big/(1 + e^{-Bt}) \tag{10-61}$$

Equation 10-61 can be rearranged to

$$-(B/C)\Big/A = 1 + e^{-Bt} \tag{10-62}$$

Subtracting one from both sides of Eq. (10-62) leads to

$$[-(B/C)\Big/A] - 1 = e^{-Bt} \tag{10-63}$$

or

$$\ln \{[-(B/C)\Big/A] - 1\} = -Bt \tag{10-64}$$

which is a linear function. If we can obtain the constant (a quotient of other constants) $B/C$, we can plot the left hand side of Eq. (10-64) as a function of $t$ and obtain $-B$ as the slope. Knowing $B$, we can obtain $C$ from the ratio of $B/C$.

In order to obtain a value for $B/C$, we return to Eq. (10-62) and note that $e^{-Bt}$ approaches zero as $t$ approaches infinity. Consequently, in the limit of very large time the equation

$$-(B/C)\Big/A = 1 \tag{10-65}$$

is true. The term "very large times" is a relative one meaning, in each case, times large enough so that the area of the bacterial colony does not increase appreciably. In the case studied here, the area does not increase appreciably after about five days. Figure 10-15 shows that the experimental curve levels off at a value of about 50 $cm^2$ at that time and the implication is that it will remain there at longer times as there is nothing in Eq. (10-61) to indicate any further increase or, indeed, change of any kind after the $e^{-Bt}$ term has gone essentially to zero. Solving Eq. (10-65) under these conditions gives the quotient $-B/C$ as 50 and permits computation of $-(B/C)\big/A$ for each of the six experimental points. This leads to the calculation of $\ln \{(-(B/C)\big/A) - 1\}$. These last two quantities are given in columns 4 and 5 of Table 10-6. (Rounding errors may be severe in these calculations.) It is now possible to plot the values of the logarithmic variation of $\{-(B/C)\big/A) - 1\}$ as a function of time as required by Eq. (10-64). The resulting slope of $-1.94$ is $-B$, which leads to a value of $C = -B/50 = -0.039$. The entire initial equation can now be written

$$dA\Big/dt = 1.94A - 0.039A^2 \tag{10-66}$$

with the solution

$$A = 50/(1 + e^{-1.94t}) \tag{10-67}$$

We see from the signs of these terms how the sigmoid curve results from Eq. (10-66), which expresses the slope of the curve of $A$ versus $t$. At first, the large value of $BA$ predominates over the small value of $CA^2$ because the constant $B$ is much larger than $C$. As $A$ becomes larger, however, it enters into the sum on the right of Eq. (10-66) as the square. Eventually, when $1.94A = -0.0388A^2$, the slope becomes zero, and the curve of $A$ versus $t$ has leveled off.

### Exercise 10-6

From Eq. (10-66), obtain the value of $A$ for which the growth curve, Fig. 10-15, levels off.

Solution 10-6. The growth curve levels off when $dA/dt = 0$. Thus,

$$1.94A - 0.0388A^2 = 0$$

Dividing by $A$,

$$1.94 = 0.0388A$$

$$A = 1.94/0.0388 = 50 \text{ cm}^2$$

## Nearly Linear Functions

Many experimental functions approach, but do not reach, linearity. Frequently early experimenters observe what they think to be linear behavior, but later, more accurate data show deviation. Nearly linear behavior may not be perfectly represented by any three-term equation, but the parabolic equation

$$y = a + bx + cx^2 \qquad (10\text{-}68)$$

is a better aproximation than a simple straight line.

Nonlinearity is generally caused by some "second order effect" apparent in the experimental data. In the example of light absorption, the simple model of a number of absorbing centers in a solution leads to the expectation that the absorption at an appropriate wavelength should vary in a linear way with concentration (Beer's law). Suppose, however, that some modification of the absorbing species takes place that is also dependent upon concentration. An example might be reaction of the abosrbing species S to form the dimer, $S_2$

$$S + S = S_2$$

By Le Chatelier's principle, increased concentration of S molecules forces the dimerization reaction to the right so that the proportion of $S_2$ relative to S increases with concentration. If the dimer absorbs light less strongly than S, the total absorption is less at higher concentrations that it "should be" on the grounds of its absorption of light at low concentration. If the dimer absorbs more strongly than S, the absorption at higher concentrations is greater than expected on the grounds of Beer's law. The system is said to deviate negatively or positively from Beer's law. If one is attempting to set up an analytical procedure to determine the amount of S in solutions using Beer's law, deviation owing to dimer formation is an annoyance, making the calculations more difficult than they would normally be. Now, instead of the simpler linear form, the absorption equation

$$A = A_0 + kc + jc^2 \tag{10-69}$$

must be used. The term $jc^2$ approximates the degree of deviation from linearity.

If, however, one is studying the equilibrium constant relating S and $S_2$, the deviation from Beer's law is just what one is looking for. In the first case, the term $jc^2$ is a correction term to be applied to a poorly-behaved function. In the second case, the magnitude of $jc^2$ is the experimental parameter needed to quantitatively characterize the equilibrium under study. In either case, one must solve Eq. (10-69) or the more general form (10-68) to obtain the best values of the parameters provided by the experimental data. There are now three terms in the approximation equation; hence there must be three least-squares parameters. The Cramer's rule treatment is correspondingly more difficult, but the principles have been treated in Chapter 9.

The sum of squares of deviations or residuals is defined as it was in the linear case except for the additional term in $x^2$

$$\sum d_i^2 = \sum r_i^2 = \sum [y_i - (a + bx_i + cx_i^2)]^2 \tag{10-70}$$

Differentiation must now be with respect to $a$, $b$, and $c$. The partial derivatives

$$\frac{\partial}{\partial b}\sum r_i^2, \ and \ \frac{\partial}{\partial c}\sum r_i^2$$

lead to

$$\sum y_i = Na + b\sum x_i + c\sum x_i^2$$
$$\sum y_i x_i = a\sum x_i + b\sum x_i^2 + c\sum x_i^3 \tag{10-71}$$

and

$$\sum y_i x_i^2 = a\sum x_i^2 + b\sum x_i^3 + c\sum x_i^4$$

which are the *three* simultaneous equations analogous to Eqs. (9-62) and (9-63).

The solution set is now $a$, $b$, and c, the nonhomogeneous vector is $\{\Sigma y_i, \Sigma y_i x_i, \Sigma y_i x_i^2\}$ and the matrix of the coefficients is

$$\begin{bmatrix} N & \sum x_i & \sum x_i^2 \\ \sum x_i & \sum x_i^2 & \sum x_i^3 \\ \sum x_i^2 & \sum x_i^3 & \sum x_i^4 \end{bmatrix} \tag{10-72}$$

By expanding the determinant corresponding to matrix (10-72), one can obtain explicit forms of $D_a$, $D_b$, $D_c$, and $D$, which are analogous to $D_b$, $D_m$, and $D$ defined in Chapter 9 [Eqs. (9-67–69). Appropriate substitution yields the solution vector $\{a, b, c\}$ as in Eqs. (9-71) and (9-72). These equations are quite cumbersome to write out; hence they will be left as exercises. We have stressed accumulation techniques in the simple least squares methods of Chapter 9. The parabolic curve-fitting program presented next, Program 10-1, is nothing more than a further elaboration, admittedly a complicated one, of those same methods.

## Program 10-1

```
10   REM SECOND DEGREE LEAST SQUARES CURVE FIT
20   DIM X(100),Y(100)
30   INPUT N
40   FOR I=1 TO N
50   INPUT  X(I),Y(I)
60   NEXT I
69   PRINT
70   PRINT "THE PROGRAM HAS ACCEPTED"N"DATA PAIRS"
71   PRINT
80   FOR I=1 TO N
81   PRINT X(I),Y(I)
82   NEXT I
85   READ S1,S2,S3,S4,S5,S6,S7
86   DATA 0,0,0,0,0,0,0
90   FOR I=1 TO N
91   S1=S1+X(I)
92   S2=S2+Y(I)
93   S3=S3+X(I)**2
94   S4=S4+X(I)**3
95   S5=S5+X(I)**4
96   S6=S6+X(I)*Y(I)
97   S7=S7+X(I)**2*Y(I)
98   NEXT I
100 D1=N*(S3*S5-S4**2)
101 D2=S1*(S1*S5-S3*S4)
102 D3=S3*(S1*S4-S3**2)
103 D=D1-D2+D3
110 A1=S2*(S3*S5-S4**2)
111 A2=S1*(S6*S5-S4*S7)
112 A3=S3*(S6*S4-S3*S7)
113 A=A1-A2+A3
120 B1=N*(S6*S5-S7*S4)
121 B2=S2*(S1*S5-S3*S4)
122 B3=S3*(S1*S7-S3*S6)
123 B=B1-B2+B3
130 C1=N*(S3*S7-S4*S6)
131 C2=S1*(S1*S7-S3*S6)
132 C3=S2*(S1*S4-S3**2)
```

```
133 C=C1-C2+C3
160 A5=A/D
170 B5=B/D
180 C5=C/D
190 PRINT A5,B5,C5
200 END

READY


RUNNH


?6
?0.,1.4994
?20.,1.4917
?40.,1.484
?60.,1.4752
?80.,1.4674
?100.,1.4752

THE PROGRAM HAS ACCEPTED 6 DATA PAIRS

0             1.4994
20            1.4917
40            1.484
60            1.4752
80            1.4674
100           1.4752
1.50122       -6.33772E-4     3.44195E-6


TIME:   0.97 SECS.
```

## Exercise 10-7

Expand the determinant (10-72) by the methods of Chapter 9 and demonstrate that it corresponds to the variable stored at location D in line 103 of Program 10-1. Expand $D_a$ by the method of Cramer's rule in Chapter 9 and demonstrate that it is the quantity stored at location A in Program 10-1. Write out A5 = A/D as in Program 10-1 so as to obtain an explicit expression for the parameter $a$ in $y = a + bx + cx^2$, the second-degree estimating equation solved by Program 10-1.

## Exercise 10-8

The refractive indices of two organic liquids are given as $Y$ in the output of Program 10-1 for values of $X$ corresponding to the percentages of A in liquid B. Write the estimating equation and use it to determine the value of the refractive index anticipated for a solution of 25% A in B.

Solution 10-8. The estimating equation is

$$Y = 1.4995 - 3.8788 \times 10^{-4} X - 1.7411 \times 10^{-7} X^2$$

For a 25% solution,

$$Y = 1.4995 - 0.0097 - 0.0001 = 1.4897$$

Application of the parabolic program offers a good test of the linearity of a data set. If the third term is very small as above ($c \cong 0$), the set is very nearly linear.

## Exercise 10-9

A stock solution of potassium dichromate plus some interfering ions was diluted to give standard solutions of the following concentrations and absorbances.

| Conc ($X$), moles/liter | Absorbance, $Y$ |
|---|---|
| $1.25 \times 10^{-3}$ | 0.673 |
| $2.50 \times 10^{-3}$ | 0.854 |
| $3.33 \times 10^{-3}$ | 1.036 |
| $4.16 \times 10^{-3}$ | 1.018 |
| $5.00 \times 10^{-3}$ | 1.009 |

Sketch the curve, determine the parabolic least squares equation to fit these points, and estimate the concentration in potassium dichromate of a solution having $A = 0.750$, assuming that the interferences are the same as in the prepared standards.

Solution 10-9. The experimental curve has scatter and shows the influence of the interfering ions clearly since it does not pass through the origin.

The parabolic least squares curve to fit these points is

$$A = 0.305 + 0.333C - 0.389C^2$$

with a standard deviation in $A$ of 0.030. Considering the roughness of the data, it is probably sufficient to obtain the concentration of the unknown solution graphically by constructing a horizontal at $A = 0.75$ until it intersects the calculated least squares parabola; then, dropping a vertical to the concentration axis, we obtain a concentration of $1.6 \times 10^{-3}$ for the unknown dichromate solution in units of moles per liter.

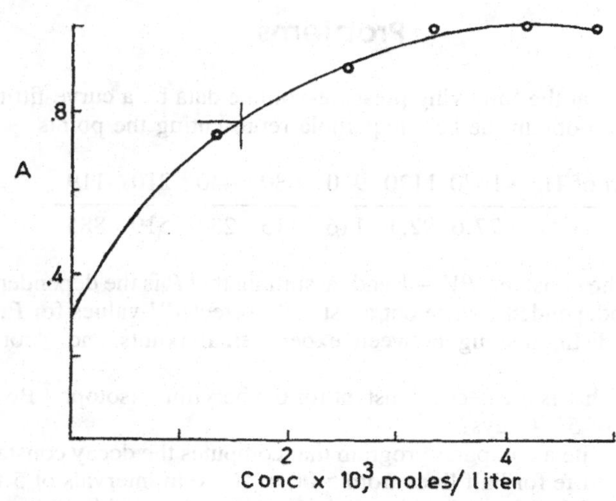


Fig. 10-16. Absorbance of potassium dichromate solutions as a function of concentration.

## Glossary

*Amylase.* An enzyme that catalyzes the conversion of starch to sugar.

*Boyle's Law.* $P = k/V$, where $P$ is the pressure of a gas, $V$ is its volume, and $k$ is a constant.

*Decay Constant.* Constant $k$ in the radioactive rate equation $dN/N = -kdt$, which is characteristic of the isotope undergoing decay.

*Doubling Time.* Time necessary for a population to double as measured from an arbitrary starting time. Constant for unlimited population growth.

*Enzyme.* Catalyst in a biological system.

*Exponential.* Function having the form $y = e^{\pm x}$.

*Half Time.* Time necessary for one half of a sample of a radioactive isotope to decay.

*Hyperbola.* Curve having the general form $(x^2/a) - (y^2/b) = 1$. A special case often encountered is $y = k/x$.

*Intrinsic Rate of Increase.* Birth rate minus death rate for a population.

*Lineweaver-Burke Plot.* Linear graph used in determining the constants in the Michaelis-Menten equation.

*Michaelis-Menten Equation.* Equation 10-27, important in enzyme kinetics.

*Parabola.* Function that may be written in the form $y = a + bx + cx^2$.

*Parameter.* Constant in an equation for given conditions that may take on a new value if the conditions change, e.g., $PV = k$, with $k$ a parameter for a given amount of gas at a given temperature.

*Second-Order Rate Equation.* Rate equation in which the sum of the exponents on the right is two.

*Sigmoid Curve.* S shaped curve resulting from a cumulative frequency distribution or limited population growth as in Fig. 10-15.

*Substrate.* Substance acted upon by an enzyme.

## Problems

*1.* Treat the following pressure-volume data by a curve-fitting technique so as to obtain the best hyperbola representing the points

| $P$, cm of Hg | 1470 | 1120 | 910 | 680 | 440 | 210 | 119 |
|---|---|---|---|---|---|---|---|
| $V$, mL | 77.6 | 92.1 | 116 | 115 | 234 | 519 | 885 |

Determine the constant, $PV = k$ and, assuming that $P$ is the dependent variable and $V$ independent, write out a list of "corrected" values for $P$. Sketch the curve, distinguishing between experimental points and "corrected" points.

*2.* What is the decay constant for the beryllium isotope $^{7}$Be that has a half time of 53.4 days?

*3.* Write a computer program that computes the decay constant from a known half-life for half lives from 5.00 to 100.0 in intervals of 5.00. Plot the curve resulting from the twenty data points generated by this program. Comment on the shape of the curve so produced. Estimate, using the curve you have drawn, the decay constants for the radioactive isotopes $^{23}$Ne, $^{24}$Na, and $^{14}$C, which have half lives of 40.0 seconds, 15.0 hours, and 5730 years, respectively.

*4.* A sample of a radioactive element had an initial count rate of $1.6 \times 10^5$ cpm. After one hour, the count rate was $1.5 \times 10^5$, after four hours, it was $1.2 \times 10^5$, after eight, cpm was $8.6 \times 10^4$, after 15 hours, cpm was $4.9 \times 10^4$ and after 20 hours, it was $3.2 \times 10^4$. Sketch the curve of ln cpm versus $t$ and the curve of cpm versus $t$. What are the decay constant and the decay half time?

*5.* Determine the rate constant for the second-order reaction

| Time, s | 30 | 60 | 120 | 240 | 480 |
|---|---|---|---|---|---|
| conc, mol/L | $8 \times 10^{-4}$ | $6.67 \times 10^{-4}$ | $5.00 \times 10^{-4}$ | $3.33 \times 10^{-4}$ | $2.00 \times 10^{-4}$ |

*6.* Plot the population curve for an insect in an environment favorable to unlimited growth and having a population doubling time of 1.35 weeks.

*7.* Determine the constants $B$ and $C$ in the equation

$$dA/dt = BA + CA^2$$

for a population of Drosophila under controlled and limited conditions

| Days | 6 | 12 | 18 | 24 | 30 | 36 | 42 | 48 |
|---|---|---|---|---|---|---|---|---|
| Population | 5 | 21 | 68 | 171 | 282 | 334 | 340 | 342 |

Plot the number of Drosophila versus time in days.

*8.* Modify Program 10-1 so that it prints out the "smoothed values" of $y$, i.e., the solutions of the least squares parabola $y = A + Bx + Cx^2$ for each input value of $x$.

*9.* Modify Program 10-1 so that it prints out the standard deviation in $y$ from the least-squares parabola $y = f(x)$.

*10.* Compute the best parabolic equation relating the solubility of sucrose in water to the temperature

| Solubility, g/L | 645 | 669 | 703 | 745 | 775 | 838 |
|---|---|---|---|---|---|---|
| Temp.,°C | 0 | 20 | 40 | 60 | 80 | 100 |

*11.* Write a program in BASIC to solve for the independent variable of a quadratic equation of the kind produced by Program 10-1 and illustrated in Exercises 10-8 and 10-9.

## Bibliography

A. J. Lotka, *Elements of Mathematical Biology*, Dover, New York, 1956.

R. W. Poole, *An Introduction to Quantitative Ecology*, McGraw-Hill, New York, 1974.

G. M. Barrow, *Physical Chemistry for the Life Sciences,* McGraw-Hill, New York, 1974.

A. Cohen, *An Elementary Treatise on Differential Equations*, Heath, New York, 1933.

C. S. Hanes, *Biochem. J.* **26**, 1406 (1932).

# Chapter 11

# Solving Simultaneous Equations

Our view of the universe is shaped by the way we look at it. Traditionally, the physical world has been treated as a system in which one independent variable acts with one dependent variable while the rest of the universe exists in an imaginary state of suspended animation. Reality is not that way of course, but it is unfair to suppose that physical scientists have failed to consider functions of many variables out of ignorance; the mathematical difficulties were simply too large to do it any other way. With the advent of the computer, systems of many variables can be studied and the impact of this technological advance will certainly be large. This chapter considers systems of many variables, in principle, indefinitely many, under the constraint that some dependent variable is a linear function of all the independent variables. As such it is an introduction, for nonlinear multivariate systems should also be studied. We will leave that for another book, however, with the comment that the principles will be the same, only the mathematical complexity will be increased.

## Simultaneous Equations in Matrix Form

Functions of many variables can be described by simultaneous equations. We have looked at some of the properties of matrices, determinants, and sets of simultaneous equations in the preceding chapters, as a means of linear and parabolic curve-fitting. In this section, we take up simultaneous equations in more detail and for their own sake.

The equations

$$\begin{aligned} 2x_1 + 3x_2 &= 8 \\ x_1 + 5x_2 &= 11 \end{aligned} \qquad (11\text{-}1)$$

lead to the solution set (1, 2), but the equations

$$2x + 3y = 8$$
$$4x + 6y = 16 \tag{11-2}$$

do not lead to a unique solution. The second pair of equations are said to be linearly dependent upon each other, but the equations in the first set are linearly independent. Whenever one equation of a set can be obtained by multiplying another equation in the same set by a constant, the equations are linearly dependent and no unique solution set exists.

The equation set

$$\sum x_{1i} + N_1 = 0$$
$$\sum x_{2i} + N_2 = 0$$
$$\vdots$$
$$\sum x_{Ni} + N_N = 0 \tag{11-3}$$

or

$$\sum x_{1i} = B_1$$
$$\sum x_{2i} = B_2$$
$$\vdots$$
$$\sum x_{Ni} = B_N$$

where $B_1 = -N_1$, $B_2 = -N_2$, etc. is nonhomogeneous; the terms not containing $x$ constitute the nonhomogeneous part. We have already seen the matrix form for the general set of $N$ equations in $N$ unknowns, Eq. (9-49). The matrix of the coefficients is frequently denoted $A$, the solution vector is called $X$, and the nonhomogeneous part, $B$.

This leads to a very economical way of writing large sets of simultaneous equations

$$AX = B \tag{11-4}$$

for the nonhomogeneous case.

Linear homogeneous sets are distinguished by the absence of terms without a variable, thus the set

$$2x - y = 0$$
$$8x - 4y = 0 \tag{11-5}$$

is a homogeneous set. It has the matrix form

$$\begin{bmatrix} 2 & -1 \\ 8 & -4 \end{bmatrix} \begin{bmatrix} x \\ y \end{bmatrix} = 0 \tag{11-6}$$

or, in general,

$$AX = 0 \tag{11-7}$$

Although sets of homogeneous equations are very interesting in some areas of theoretical physics and chemistry, nonhomogeneous sets are more interesting to the experimentalist; hence we shall confine ourselves to *nonhomogeneous, independent* sets of linear simultaneous equations.

## Matrix Algebra

As in any algebra, we start by defining some elementary operations: addition, subtraction, multiplication, and division. Along the way we shall develop some new terminology appropriate to describing matrices.

Matrices are added merely by adding each element of one matrix to the corresponding element of the other. An *element* will be a number in most of what follows, but there are many situations in which the elements of a matrix are not simple numbers. They may be sums, as we have already seen, or they may be more complicated mathematical entities such as polynomials, integrals, etc. Elements are denoted by their position within the matrix according to row (horizontal) and column (vertical). The elements $a_{43}$ and $a_{79}$ occupy unique positions, the fourth row, third column in the former case and the seventh row, ninth column in the latter.

With these rules in mind, matrix addition is very simple, e.g.,

$$\begin{bmatrix} 2 & 3 \\ 1 & 5 \end{bmatrix} + \begin{bmatrix} -1 & 1 \\ 3 & -2 \end{bmatrix} = \begin{bmatrix} 1 & 4 \\ 4 & 3 \end{bmatrix} \tag{11-8}$$

but if the matrices are large enough or if many of them need to be added, addition by hand may be sufficiently tedious that a computer program would be useful. In some computer languages, the job might be complicated enough to discourage writing a program for simple addition, but in BASIC, system subroutines have been written to perform matrix manipulations via the MAT commands, of which addition offers the simplest example. Program 11-2 adds matrices via the MAT command, but as the commentary to Program 11-1 shows, the matrices have to be entered in a specific way; hence attention to this program and running it for simple matrices will provide a good background for later matrix programs. Since we are interested in $N$ equations in $N$ unknowns, we need only consider square matrices, $N$ on an edge, for the time being.

## Program 11-1

Subtraction of matrices is the inverse of addition and should require no further comment.

```
10   DIM A(3,4)
20   MATREAD A
30   DATA 1,0,-6,2,-1,1,-5,4,0,3,1,8
40   MATPRINT A
50   END
RUNNH


 1              0              -6              2

-1              1              -5              4

 0              3               1              8


TIME:   0.30 SECS.
```

**Commentary on Program 11-1.** Program 11-1 does nothing but read and write a matrix; however, there are enough new factors involved to warrant a single program as a preliminary to the programs following that perform operations on matrices. The first statement dimensions the matrix *A* just as a dimension statement preceded programs using subscripted variables. However, there must be two numbers in the dimension statement for a matrix, one for the number of rows and the second for the number of columns. In this case, *A* is dimensioned as a rectangular $3 \times 4$ matrix. Statement 20 is one of a number of very convenient MAT statements in BASIC. Each of these statements calls up a system subroutine which performs some operation on a matrix or two matrices. In languages lacking the MAT statements, these subroutines must be written by the programmer and can be rather complicated. Statement 20 causes the matrix *A* to be read into memory from the data array in statement 30. Each number in the DATA array is read, in sequence. If a number is out of sequence in the DATA array, it will be out of place in the matrix. Statement 40 prints the matrix in its proper form as governed by the DIM statement.

Matrix multiplication is not as obvious as matrix addition. The rule for multiplying matrices, AB is to sum the products of the first row of *A* with the first column of *B* to obtain the element $c_{11}$ of the product matrix *C*, sum the products of the first row of *A* with the second column of *B* for $c_{12}$, with the third for $c_{13}$, etc. When the first row of *A* has been treated as above for all columns of *B*, the first row of *C* will have been obtained. The second row of *C* is produced by repeating the routine using the second row of *A* and all columns of *B*. This is continued until all elements of *C* have been generated. The routine is demonstrated here for the two by two case

$$\begin{bmatrix} a_{11} & a_{12} \\ a_{21} & a_{22} \end{bmatrix} \begin{bmatrix} b_{11} & b_{12} \\ b_{21} & b_{22} \end{bmatrix}$$

$$= \begin{bmatrix} a_{11}b_{11} + a_{12}b_{21} & a_{11}b_{12} + a_{12}b_{22} \\ a_{11}b_{11} + a_{12}b_{21} & a_{11}b_{12} + a_{12}b_{22} \end{bmatrix} \qquad (11\text{-}9)$$

It is easy to see that for a matrix of any size, a computer routine would be appropriate. BASIC provides a subroutine for matrix multiplication, as well as the one for addition, which is also shown in Program 11-2. Check all results by hand calculation.

## Program 11-2

```
10  DIM A(3,3),B(3,3),C(3,3),D(3,3),E(3,3)
20  MATREAD A
30  MATREAD B
40  PRINT"THE TWO ORIGINAL  MATRICES ARE:"
45  PRINT"MATRIX A:"
50  MATPRINT A
51  PRINT
52  PRINT"MATRIX B:"
55  MATPRINT B
60  MAT C=A+B
70  MAT D=A-B
80  MAT E=A*B
85  PRINT
90  PRINT"THE SUM OF MATRICES A AND B IS:"
100 MATPRINT C
105 PRINT
110 PRINT"THE DIFFERANCE BETWEEN MATRICES A AND B IS:"
120 MATPRINT D
125 PRINT
130 PRINT"THE PRODUCT OF MATRICES A AND B IS:"
140 MATPRINT E
145 DATA 4,2,1,0,6,1,1,2,-2,-4,8,0,0,2,1,1,-4,-5
150 END

READY

RUNNH

THE TWO ORIGINAL  MATRICES ARE:
MATRIX A:

 4              2              1

 0              6              1

 1              2             -2


MATRIX B:

-4              8              0

 0              2              1

 1             -4             -5
```

```
THE SUM OF MATRICES A AND B IS:

0              10             1

0              8              2

2              -2             -7

THE DIFFERANCE BETWEEN MATRICES A AND B IS:

8              -6             1

0              4              0

0              6              3

THE PRODUCT OF MATRICES A AND B IS:

-15            32             -3

 1             8              1

-6             20             12


TIME:  1.02 SECS.
```

Commentary on Program 11-2. After dimensioning, reading, and printing input matrices $A$ and $B$ via MATREAD and MATPRINT statements, Program 11-2 performs the actual operations described in the previous paragraphs in steps 60, 70, and 80, MAT add, subtract, and multiply statements. Following these operations, standard MATPRINT and DATA statements complete the program.

Matrix division is not defined, but multiplication by an *inverse* matrix is. Just as multiplying a number by its inverse, say 5 times 1/5 equals 1, multiplying a matrix, $A$, by its inverse, $A^{-1}$, gives the unit matrix, that matrix with ones on the principal diagonal and zeros elsewhere, e.g.,

$$\begin{bmatrix} 1 & 0 & 0 \\ 0 & 1 & 0 \\ 0 & 0 & 1 \end{bmatrix} \tag{11-10}$$

for the $3 \times 3$ case.

Using ordinary numbers, we know that, if 1/5 is the inverse of 5, then 5 is the inverse of 1/5. The same holds for matrices. If $A^{-1} = B$ then $B^{-1} = A$.

**Exercise 11-1**

If

$$A = \begin{bmatrix} 0 & -1 \\ 1 & 0 \end{bmatrix}$$

demonstrate that

$$\begin{bmatrix} 0 & 1 \\ -1 & 0 \end{bmatrix}$$

is its inverse, $A, ^{-1}$.

Solution 11-1. If the matrices are, in fact, inverses of each other, then their product should be the unit matrix. Multiplying,

$$\begin{bmatrix} 0 & -1 \\ 1 & 0 \end{bmatrix} \begin{bmatrix} 0 & 1 \\ -1 & 0 \end{bmatrix} = \begin{bmatrix} 0+1 & 0+0 \\ 0+0 & 1+0 \end{bmatrix} = \begin{bmatrix} 1 & 0 \\ 0 & 1 \end{bmatrix}$$

Hence the matrices given are inverses of each other.

The central problem in what is to follow is the process of determining the inverse of a known matrix. The process of finding $A^{-1}$ from $A$ is known as *matrix inversion*. One way of doing this is by row transformations. In this method, one multiplies any row in matrix $A$ and adds it to or subtracts it from any other row with the ultimate end of transforming $A$ into the unit matrix. (Merely multiplying a row by a constant and omitting the addition step is also a legitimate row transformation.) For square matrices of the coefficients of linearly independent sets of nonhomogeneous equations, the kind we have restricted ourselves to, this can always be done.

If one applies the same sequence of row transformations to the unit matrix, it will be converted to $A^{-1}$. For convenience, $A$ and the unit matrix are often set side by side and the conversions are performed at the same time.

Starting with the matrix of the coefficients of the Eq. set (11-1) and the unit matrix,

$$\begin{bmatrix} 2 & 3 \\ 1 & 5 \end{bmatrix} \begin{bmatrix} 1 & 0 \\ 0 & 1 \end{bmatrix} \tag{11-11}$$

we may multiply the bottom row of both by 1 (i.e., leave it unchanged) and subtract the result from the top row. This is a typical row transformation leading to

$$\begin{bmatrix} 1 & -2 \\ 1 & 5 \end{bmatrix} \begin{bmatrix} 1 & -1 \\ 0 & 1 \end{bmatrix} \tag{11-12}$$

Subtracting the top row from the bottom row yields

$$\begin{bmatrix} 1 & -2 \\ 0 & 7 \end{bmatrix} \begin{bmatrix} 1 & -1 \\ -1 & 2 \end{bmatrix} \tag{11-13}$$

Now divide the bottom row by 7. This is still a row transformation even though we do not add the result to or subtract the result from the top row. We have

$$\begin{bmatrix} 1 & -2 \\ 0 & 1 \end{bmatrix} \begin{bmatrix} 1 & -1 \\ -1/7 & 2/7 \end{bmatrix} \tag{11-14}$$

Multiplying the bottom row by 2 and adding the result to the top row yields

$$\begin{bmatrix} 1 & 0 \\ 0 & 1 \end{bmatrix} \begin{bmatrix} 5/7 & -3/7 \\ -1/7 & 2/7 \end{bmatrix} \tag{11-15}$$

We have successfully converted $A$ on the left to the unit matrix by one set of row transformations. It should be mentioned in passing that there are other sequences of simple row transformations that would have done just as well. The sequence of steps was simultaneously applied to the unit matrix adjacent to $A$ on the right with the purpose of converting it to $A^{-1}$. To test this, we multiply $A$ and what we suppose to be $A^{-1}$

$$\begin{bmatrix} 2 & 3 \\ 1 & 5 \end{bmatrix} \begin{bmatrix} 5/7 & -3/7 \\ -1/7 & 2/7 \end{bmatrix}$$

$$= \begin{bmatrix} 10/7 - 3/7 & -6/7 + 6/7 \\ 5/7 - 5/7 & -3/7 + 10/7 \end{bmatrix} = \begin{bmatrix} 1 & 0 \\ 0 & 1 \end{bmatrix} \tag{11-16}$$

We conclude, it is indeed true that

$$A^{-1} = \begin{bmatrix} 5/7 & -3/7 \\ -1/7 & 2/7 \end{bmatrix} \tag{11-17}$$

It is clear that inversion of a large matrix can be very laborious, hence one of the most welcome amenities of the BASIC language is the MAT INV subroutine that does it all and is illustrated in Program 11-3. This program includes multiplication of the supposed $A^{-1}$, which has been stored at location $Z$ into $A$, to ascertain whether the product is the unit matrix.

## Program 11-3

```
10  DIM A(3,3),I(3,3),B(3,3),Z(3,3)
20  MATREAD A
30  MAT B=A
40  MATZ=INV(B)
50  PRINT "THE ORIGINAL MATRIX (A) IS:"
60  MATPRINT A
70  PRINT "THE INVERSE OF (A) IS:"
80  MATPRINT Z
90  MATI=Z*A
100 PRINT "THE IDENTITY MATRIX (I) IS:"
110 MATPRINT I
120 DATA 1,1,2,3,0,0,4,-2,6
130 END
READY
RUNNH

THE ORIGINAL MATRIX (A) IS:

 1              1             2

 3              0             0

 4             -2             6
THE INVERSE OF (A) IS:

 0              0.333333      0

 0.6            6.66667E-2   -0.2

 0.2           -0.2           0.1

THE IDENTITY MATRIX (I) IS:

 1              0             0

 0              1             0

-3.72529E-9     0             1


TIME:  0.62 SECS.
```

**Commentary on Program 11-3.** After dimensioning four 3 × 3 matrices, $A$ is read in and stored in two memory locations under the names $A$ and $B$, inverted and stored as the matrix $Z$. $A$ and $Z$ are then printed out with appropriate captions. Statement 90 is the MAT subroutine for multiplying matrices and the result of $AZ$ is stored as matrix $I$. If $Z$ is in fact the inverse of $A$,

$$AZ = AA^{-1} = I \tag{11-18}$$

the result should be the identity matrix

$$I = \begin{bmatrix} 1 & 0 & 0 \\ 0 & 1 & 0 \\ 0 & 0 & 1 \end{bmatrix} \tag{11-19}$$

and it is, almost exactly. The off diagonal elements are not always output as zero, but may be very small numbers. These small residuals are the result of truncation errors in the process of matrix inversion and multiplication. It is not uncommon to observe a very small residual where a zero should be, or a value of 0.99999999 where in the longhand calculation, a one should appear.

## Simultaneous Equations by Matrix Inversion

Now that we have some grounding in matrix algebra, we know how to input a matrix in BASIC and we have the statements to invert and multiply matrices, we are able to develop a general method for solving nonhomogeneous, linear, independent simultaneous equations. The matrix equation

$$AX = B \tag{11-20}$$

is premultiplied on both sides by $A^{-1}$ yielding

$$A^{-1}AX = A^{-1}B \tag{11-21}$$

but $A^{-1}A$ is the unit matrix, which functions as the number one does in ordinary algebra; mutiplication of $X$ by one in either ordinary algebra or matrix algebra leaves it unchanged. Hence

$$X = A^{-1}B \tag{11-22}$$

i.e., the solution vector of a set of simultaneous equations is nothing more than the product of the inverse of the coefficient matrix times the nonhomogeneous vector. Program 11-4 solves the set

$$\begin{aligned} 3x + y &= 5 \\ x + y &= 3 \end{aligned} \tag{11-23}$$

to give the obvious solution set (1, 2).

## Program 11-4

```
10   DIM A(2,2),X(2,1),B(2,1),Z(2,2)
20   MATREAD A
21   DATA 3,1,1,1
30   MATREAD B
31   DATA 5,3
40   MATZ=INV(A)
50   MATX=Z*B
60   MATPRINT X
70   END

READY
RUNNH


 1.

 2.


TIME:   0.15 SECS.
```

## Exercise 11-2

Solve the sets of simultaneous equations

$$2x + y = 8$$

$$x + 5y = 11$$

and

$$x_1 + 2x_2 - x_3 + x_4 = 2$$

$$x_1 - 2x_2 + x_3 - 3x_4 = 6$$

$$2x_1 - x_2 + 2x_3 + x_4 = 4$$

$$3x_1 + 3x_2 + x_3 - 2x_4 = 10$$

Solution 11-2

```
READY
21   DATA 2,1,1,5
31   DATA 8,11
RUNNH


 3.22222

 1.55556
```

```
TIME:   0.16 SECS.

READY
10   DIM A(4,4),X(4,1),B(4,1),Z(4,4)
21   DATA 1,2,-1,1,1,-2,1,-3,2,1,2,1,3,3,1,-2
31   DATA 2,6,-4,10
RUNNH



 1.

 1.

-2.

-3



TIME:   0.22 SECS.
```

## Simultaneous Analysis

Not infrequently, two substances are present in a solution, both of which are present in unknown concentration and both of which absorb light in the same region of the spectrum. We shall take as our example two hypothetical metal complexes, *X* and *Y*, that absorb in the visible region. Their spectra are shown in Fig. 11-1.

Figure 11-1 shows the spectra of the complexes at arbitrary concentrations [*X*] and [*Y*] along with the spectrum of a solution that has been

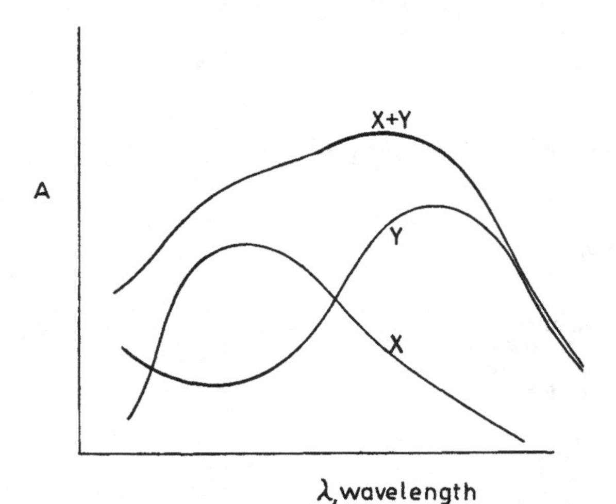


Fig. 11-1. Absorption spectra of complexes *X* and *Y* and a mixture of *X* and *Y*.

made up to the same concentration in both $X$ and $Y$. Each point on the spectrum for the mixture is the sum of the absorbance owing to component $X$ plus that owing to $Y$ at any given wavelength. The spectra are said to be additive. If Beer's law is assumed to hold for each component in the mixture, $A = abc$ for each component. Taking $b = 1.00$ for convenience, and considering all components in the mixture,

$$A = \sum a_i c_i \quad (11\text{-}24)$$

where $a_i$ is the absorptivity at any specified wavelength, $A$ is the absorbance and $c_i$ is the concentration of the ith component in solution.

Equation (11-24) provides for simultaneous determination of unknowns by spectrophotometry. If, for the simple case of two unknowns that have been qualitatively identified, we measure $A$ and $A'$ for pure samples of known concentrations at two different wavelengths, $\lambda$ and $\lambda'$, we can solve for $a = A/c$ and $a' = A'/c$ at wavelengths $\lambda$ and $\lambda'$, respectively. Let the determination be carried out for the first component of the mixture yielding absorptivity values $a_1$ and $a_1'$ followed by a similar determination of a sample of pure component 2 of known concentration to yield $a_2$ and $a_2'$.

$$A_{\text{sum}} = a_1 c_1 + a_2 c_2 \quad (11\text{-}25)$$

for the mixture measured at $\lambda$ and

$$A'_{\text{sum}} = a_1' c_1 + a_2' c_2 \quad (11\text{-}26)$$

for the same mixture measured at $\lambda'$ in a cell having the same light path. Simultaneous solution of these two equations leads to the desired concentrations $c_1$ and $c_2$.

## Exercise 11-3

Tyrosine and tryptophan have absorptivities of approximately $5.0 \times 10^3$ and $2.0 \times 10^3$ L cm$^{-1}$ mol$^{-1}$ at 250 nm in the ultraviolet and $1.5 \times 10^3$ and $5.0 \times 10^3$ at 275 nm. If a solution containing both has absorbance values measured in a 1.00 cm cell of $A = 0.582$ and $A' = 0.465$ at the lower and higher wavelengths, respectively, what is the concentration of each?

Solution 11-3. Let the concentration of tyrosine be $c_1$ and that of tryptophan be $c_2$. At 250 nm,

$$A = 0.582 = 5.0 \times 10^3 c_1 + 2.0 \times 10^3 c_2$$

and at 275 nm,

$$A' = 0.465 = 1.5 \times 10^3 c_1 + 5.0 \times 10^3 c_2$$

The solution set is $9.0 \times 10^{-5}$ and $6.6 \times 10^{-5}$, the concentrations of tyrosine and tryptophan, respectively, in units of moles per liter.

The previous method is by no means restricted to two component mixtures. In a well-known paper on mass spectroscopy, Gifford, Rock, and Commaford studied mixtures of *n*-butyl, *sec*-butyl, *t*-butyl and isobutyl alcohols by mass spectroscopy under conditions such that each component contributed to distinguishable mass spectroscopic peaks in proportion to the amount present. Work up of the data leads to the equations

$$\begin{aligned} 0.26x_1 + 100.0x_2 + 17.78x_3 + 4.98x_4 &= 301.5 \\ 90.58x_1 + 1.47x_2 + 1.02x_3 + 2.46x_4 &= 126.7 \\ 6.59x_1 + 0.59x_2 + 100.00x_3 + 5.03x_4 &= 322.6 \\ 0.79x_1 + 0 + 0.29x_3 + 9.06x_4 &= 14.8 \end{aligned} \tag{11-27}$$

where $x_1$, $x_2$, $x_3$, and $x_4$ are the percentages of the four alcohols named above in that order.

### Exercise 11-4

Show that the previous set of simultaneous equations leads to the following mole % for the four alcohols named above: 24.4, 25.0, 24.8, 25.8.

### Exercise 11-5

A series of potassium dichromate solutions were made up containing 21.0, 53.0, 64.0, 85.0, and 106 parts per million (ppm) of chromium. A similar series of potassium permanganate solutions were made up containing 3.45, 6.90, 10.4, 13.8, 17.3 ppm of manganese. The absorbance of each of these ten solutions was measured at two wavelengths, $\lambda = 440$ and $\lambda' = 525$ nm. The absorbances were

$\lambda$(Cr) 0.099, 0.241, 0.289, 0.371, 0.456
$\lambda'$(Cr) 0.009, 0.023, 0.025, 0.029, 0.042
$\lambda$(Mn) 0.010, 0.029, 0.031, 0.040, 0.049
$\lambda'$(Mn) 0.148, 0.289, 0.441, 0.580, 0.740

Compute the coefficient matrix for simultaneous determination of dichromate and permanganate.

Solution 11-5. The slope of $A$ as a function of concentration is determined by repeated application of Program 9-1 or its equivalent. The slopes are

$$\lambda(\text{Cr}): a_1 = 4.3871 \times 10^{-3}$$

$$\lambda'(\text{Cr}): a_2 = 3.846 \times 10^{-4}$$

$$\lambda(\text{Mn}): a_3 = 2.9756 \times 10^{-3}$$

$$\lambda'(\text{Mn}): a_4 = 4.242 \times 10^{-2}$$

The equations we wish to solve are

$$a_1(\text{Cr}) + a_3(\text{Mn}) = A \text{ at } 440 \text{ nm}$$

$$a_2(\text{Cr}) + a_4(\text{Mn}) = A' \text{ at } 525 \text{ nm}$$

hence, rounding to a proper number of significant figures, the coefficient matrix is

$$\begin{bmatrix} 4.39 \times 10^{-3} & 2.98 \times 10^{-3} \\ 3.85 \times 10^{-4} & 4.24 \times 10^{-2} \end{bmatrix}$$

## Exercise 11-6

Three solutions containing both potassium dichromate and potassium permanganate were investigated under identical conditions to those in the standardization procedure described in Exercise 11-5. The absorbances were $A = 0.421, 0.232, 0.112$, and $A' = 0.120, 0.440, 0.645$. What were the concentrations of Cr and Mn in each of the three solutions?

Solution 11-6. Since conditions were identical to those in Exercise 11-5 (in particular, the light path was the same) we may apply the coefficient matrix that we obtained there. The first pair of absorbance values, $A = 0.421$ and $A' = 0.120$ constitutes the nonhomogeneous vector for the first determination, (0.232, 0.440), is the second nonhomogeneous vector and (0.112, 0.645) is the third. When the coefficient matrix is entered as the 21 DATA statement in Program 11-4 and the first nonhomogeneous vector is entered as 31 DATA, the solution set is

$$(\text{Cr}, \text{Mn}) = (94.6, 1.97)$$

The second and third nonhomogeneous vectors lead to

$$(\text{Cr}, \text{Mn}) = (46.1, 9.96)$$

$$(\text{Cr}, \text{Mn}) = (15.1, 15.1)$$

where all concentrations are in ppm.

## Multiple Regression

Sometimes it is not possible to study the effect of one independent variable in the absence of the other. Then, obtaining the coefficient matrix for simultaneous equation solving cannot be done by simple linear least squares, as it was in Exercise 11-5. An extended mathematical method is needed, and since the result of a linear least squares treatment is often called a *regression line*, the new technique involving, as it does, two or more independent variables, is called *multiple regression*. We shall give the theory of multiple regression then show how the method can be used to obtain the coefficient matrix in Solution 11-5.

The necessary constants, $a_{11}$ and $a_{12}$ can be obtained by observing the absorbance for two different concentration ratios $C_1/C_2$ and $C_1'/C_2'$

$$a_{11}C_1 + a_{12}C_2 = A_1$$
$$a_{11}C_1' + a_{12}C_2' = A_1' \tag{11-28}$$

i.e., ratios such that the second equation in the set is not a linear combination of the first. Repetition of this pair of measurements at the second wavelength leads to absorbancies $a_{21}$ and $a_{22}$

$$a_{21}C_1 + a_{22}C_2 = A_2$$
$$a_{21}C_1' + a_{22}C_2' = A_2' \tag{11-29}$$

The mathematical requirements for unique determination of $n$ independent unknowns (in this case 2) is that there be $n$ measurements made on the system so as to provide $n$ independent equations. In practice, however, because of experimental error, this is a minimum requirement and may be expected to lead to the least accurate solution set for the system just as establishing the slope of a straight line through the origin by one experimental point may be expected to yield the least accurate slope, inferior to that obtained from 2, 3, or $p$ experimental points. Accepted experimental practice dictates that we obtain many experimental points for a function of one variable and determine the "best" straight line through them by means of a least squares regression procedure. The analogous procedure for a function of two or more variables is to obtain many equations and extract the "best" parameters, $a_{11}$, $a_{12}$. . .$a_{21}$, $a_{22}$. . .$a_{nn}$, from them by an averaging procedure. These parameters, written as an $n \times n$ array, are the coefficient matrix

$$\begin{bmatrix} a_{11}a_{12} & \cdots & a_{1n} \\ a_{21}a_{22} & \cdots & a_{2n} \\ \cdots & \cdots & \cdots \\ a_{n1}a_{n2} & \cdots & a_{nn} \end{bmatrix} \tag{11-30}$$

Optimal solution for $n$ unknowns requires that the coefficient matrix be obtained from an $n \times p$ input matrix where $p$, the number of rows of input data, is larger than $n$, preferably much larger. When there are more than the minimum number of equations from which the coefficient matrix is to be extracted, we shall refer to the coefficient matrix as having been *overdetermined*. Clearly, $n$ equations can be selected from among the $p$ available equations, but this is precisely what we do not wish to do because we must subjectively discard some of the experimental data which may have been gained at considerable expense in time and money.

## Least Squares Treatment of Multiple Regression

The function $y = m_1x_1 + m_2x_2$ describes a plane. Let us restrict it to positive values of $x_1$, $x_2$, and $y$. One measurement of $y$ yields

$$y_1 = m_1x_{11} + m_2x_{12} \tag{11-31}$$

and a second measurement yields

$$y_2 = m_1x_{21} + m_2x_{22} \tag{11-32}$$

Equations (11-31) and (11-32) can be written matrix form

$$XM = Y \tag{11-33}$$

where $X$ is the input matrix of independent variables

$$X = \begin{bmatrix} x_{11}x_{12} \\ x_{21}x_{22} \end{bmatrix} \tag{11-34}$$

$Y$ is the nonhomogeneous vector of dependent variables, and $M$ is the solution vector.

Suppose the independent variables are measurable with an indefinite degree of accuracy, but that $y$ suffers random error. Suppose further that we wish to determine $m_1$ and $m_2$ from $y_1$ and $y_2$ using a least squares procedure. The deviation of $y_1$ from its least squares value is

$$d_1 = m_1x_{11} + m_2x_{12} - y_1 \tag{11-35}$$

and the deviation of $y_2$ is

$$d_2 = m_1x_{21} + m_2x_{22} - y_2 \tag{11-36}$$

The sum of squares is

$$d_1^2 + d_2^2 = (m_1x_{11} + m_2x_{12} - y_1)^2 + (m_1x_{21} + m_2x_{22} - y_2)^2 \tag{11-37}$$

Minimizing with respect to $m_1$,

$$\partial\sum d^2/\partial m_1 = 2(m_1x_{11} + m_2x_{12} - y_1)x_{11} + 2(m_1x_{21} + m_2x_{22} - y_2)x_{21} = 0 \tag{11-38}$$

and with respect to $m_2$,

$$\partial \sum d^2/\partial m_2 = 2(m_1x_{11} + m_2x_{12} - y_1)x_{12} + 2(m_1x_{21} + m_2x_{22} - y_2)x_{22} = 0 \tag{11-39}$$

Now

$$m_1x_{11}^2 + m_2x_{12}x_{11} + m_1x_{21} + m_2x_{22}x_{21} = y_1x_{11} + y_2x_{21} \tag{11-40}$$

and

$$m_1x_{11}x_{12} + m_2x_{12}^2 + m_1x_{21}x_{22} + m_2x_{22}^2 = y_1x_{12} + y_2x_{22} \tag{11-41}$$

These simultaneous equations are the normal equations for the problem. Notice the simpler set of terms on the right. The values of $x$ and $y$ commute (that is, $xy = yx$); hence it is equivalent to

$$\begin{array}{c} x_{11}y_1 + x_{21}y_2 \\ x_{12}y_1 + x_{22}y_2 \end{array} \tag{11-42}$$

which can be written in matrix form

$$\begin{bmatrix} x_{11}x_{21} \\ x_{12}x_{22} \end{bmatrix} \begin{bmatrix} y_1 \\ y_2 \end{bmatrix} \tag{11-43}$$

The matrix in $x$ is the matrix we get by exchanging all elements of the input matrix not on the principal diagonal. It is called the transpose of the original matrix $X$, denoted $X^{TR}$. We may write the right-hand part of the normal equations

$$X^{TR}Y \tag{11-44}$$

Following the method we used on the right-hand side of Eq. (11-40) and (11-41), we can rearrange terms on the left to obtain

$$\begin{array}{c} x_{11}x_{11}m_1 + x_{11}x_{12}m_2 + x_{21}x_{21}m_1 + x_{21}x_{22}m_2 \\ x_{12}x_{11}m_1 + x_{12}x_{12}m_2 + x_{22}x_{21}m_1 + x_{22}x_{22}m_2 \end{array} \tag{11-45}$$

The appearance of two $x$ values and one $m$ in each term in each sum makes it apparent that the $m$ vector will have to be multiplied by an $X$ matrix twice. The arrangement of the subscripts makes it clear that one of the matrices is $X$ and the other $X^{TR}$. In fact

$$X^{TR}XM \tag{11-46}$$

reproduces the vector of sums above. Two very important facts emerge here. One is obvious: although this demonstration was worked out for the simple 2 × 2 case, there is no limit on size; the demonstration is exactly the same for the $n \times n$ case. The second fact may not be obvious, but is perhaps more important: *the input matrix need not be square*. By the geometric properties of any rectangular matrix, it may always be multiplied into its own transpose. The result is a square matrix of the smaller of the two dimensions of the input rectangular matrix. Indeed, for the present treatment to be nontrivial, the input matrix *must* be rectangular; a square input matrix with $X^{-1}$ defined reduces to the problem of $n$ equations, which, as we said in the introduction to this section, is not the problem we want to solve. From this point on, envision $X$ as an $n$ by $p$ matrix with $p > n$.

Recall that normal equations arising from the least squares technique are simultaneous equations. The normal equations in matrix form,

$$X^{TR}XM = X^{TR}Y \tag{11-47}$$

are often written

$$(X^{TR}X)M = Q \tag{11-48}$$

where $(X^{TR}X)$ emerges as the coefficient matrix, $Q$ the nonhomogeneous vector, and $M$ the solution vector consisting of an $n$-fold vector of slopes, $m_i$. Now, solution for the least squares value of the slope vector, $M$, follows by inverting the matrix $(X^{TR}X)$ and premultiplying it into both sides of Eq. (11-48)

$$(X^{TR}X)^{-1}(X^{TR}X)M = (X^{TR}X)^{-1}Q \tag{11-49}$$

or

$$M = (X^{TR}X)^{-1}Q \tag{11-50}$$

The sample variance, $S^2$, for each solution vector may be calculated from the squares of the residuals in the usual way

$$S^2 = \frac{1}{p-n-1} \sum R^2 \tag{11-51}$$

Student's $t$-tests of null hypotheses can be carried out and confidence limits of each of the $n$ slopes can be calculated for the $n$-variate problem. Multiple correlation coefficients can be obtained for the relationship between the dependent variable and each of the independent variables. A more advanced text on regresion analysis (Chatterjee, 1977) treats these problems in both matrix and extended algebraic notation.

Each independent nonhomogeneous vector leads to a new solution vector $M$, each with $n$ components. If the process just described is repeated $n$ times, an $n \times n$ coefficient matrix is generated that can be used to solve for independent variables, $x_1$, $x_2 \ldots x_n$, using new nonhomogeneous vectors as the input data. The process is equivalent to the univariate method of establishing the slope of a linear calibration curve using a number of known values of $x$ followed by determination of unknown values of

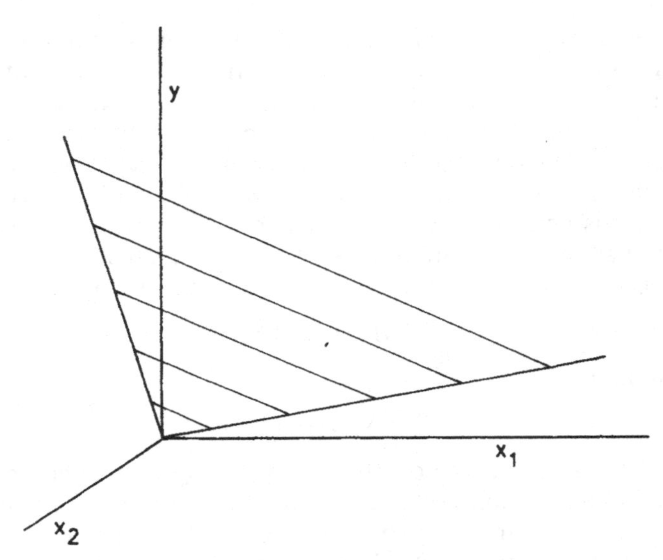


Fig. 11-2. Calibration hypersurface for $y = f(x_1, x_2)$.

$x$ from the established curve. The $n \times n$ slope matrix has established a calibration hypersurface in $n$-dimensional space for the simultaneous determination of $n$ unknowns using an $n$-fold input vector. The two-dimensional calibration plane for $y = f(x_1, x_2)$ is shown in Fig. 11-2. There is no reason, however, to restrict oneself to hyperplanes because the same method can be extended to functions that are not linear in $x$ merely by letting the $x_i$ equal functions of a higher degree. Also, we have assumed the plane passes through the origin for simplicity. This is not a necessary restriction.

## Program 11-5

The following program performs all the matrix operations just described and prints out the coefficient vector corresponding to any nonhomogeneous input vector for a given overdetermined matrix of independent variables, for example, concentrations of Mn and Cr ions like those in Example 11-5, and the corresponding absorbances. In modification 11-5a, the intermediate steps are printed out with explanatory notes to make the functioning of the program and of the preceeding theoretical discussion easier to follow. To modify this program for application to a new problem, it is not only necessary to change the data statements, but to make any necessary changes in the dimension statements so that they conform exactly to the dimensions of the new input matrix. Note that vectors are dimensioned as though they were one-column matrices.

```
10  DIM A(5,2),B(2,5),C(2,2),D(2,2),E(2,1),Y(5,1)
15  MATREAD A
20  DATA 53,8.65,27,12.98,80,4.33,0,17.3,106,0
40  MATB=TRN(A)
60  MATC=B*A
80  MATREAD Y
90  DATA .251,.149,.361,.049,.456
110 MATD=INV(C)
130 MATE=B*Y
150 MATF=D*E
155 PRINT'THE COLUMN VECTOR OF THE COEFFICIENTS IS'
160 MATPRINT F
170 END

READY
RUNNH


THE COLUMN VECTOR OF THE COEFFICIENTS IS


 4.31636E-3

 2.72431E-3


TIME:   0.35 SECS.
```

**Commentary on Program 11-5.** Seven matrices must be dimensioned in this program even though only one matrix and one vector are input. Note that the vertical dimension of the matrix is the first number and the horizontal dimension is the second number in the dimension statements. Vectors are regarded as one-dimensional matrices and there are several matrices that need to be computed and stored as intermediate steps in the calculation. The transpose of the input matrix is obtained by the simple rearrangement implicit in the matrix statement MAT B = TRN(A). The product of A and its transpose is obtained at statement 60 and multiplied into the nonhomogeneous vector at 130. Finally, the inverse of A*B is multiplied into B*Y to yield the desired column vector of the coefficient matrix. This procedure is repeated with a second nonhomogeneous vector in the following exercise to obtain the entire coefficient matrix. In the general $n \times n$ case the procedure would be repeated $n$ times.

## Program 11-5A

```
10  DIM A(5,2),B(2,5),C(2,2),D(2,2),E(2,1),Y(5,1)
15  MATREAD A
20  DATA 53,8.65,27,12.98,80,4.33,0,17.3,106,0
21  PRINT'THE INPUT MATRIX,A, IS'
23  MATPRINT A
40  MATB=TRN(A)
41  PRINT'THE TRANSPOSE OF THE INPUT MATRIX, B, IS'
43  MATPRINT B
```

```
60  MATC=B*A
61  PRINT"THEIR PRODUCT, C, IS"
63  MATPRINT C
80  MATREAD Y
90  DATA .251,.149,.361,.049,.456
91  PRINT"THE NONHOMOGENEOUS VECTOR IS"
93  MATPRINT Y
110 MATD=INV(C)
111 PRINT"THE INVERSE OF C IS"
113 MATPRINT D
130 MATE=B*Y
131 PRINT"THE PRODUCT OF THE NONHOMOGENEOUS VECTOR AND C INVERSE IS"
133 MATPRINT E
150 MATF=D*E
155 PRINT"THE COLUMN VECTOR OF THE COEFFICIENTS IS"
160 MATPRINT F
170 END

READY


THE INPUT MATRIX,A, IS


 53             8.65

 27             12.98

 80             4.33

 0              17.3

 106            0

THE TRANSPOSE OF THE INPUT MATRIX, B, IS


 53             27             80             0              106

 8.65           12.98          4.33           17.3           0

THEIR PRODUCT, C, IS


 21174          1155.31

 1155.31        561.342

THE NONHOMOGENEOUS VECTOR IS


 0.251

 0.149

 0.361

 4.90000E-2

 0.456

THE INVERSE OF C IS


 5.32022E-5    -1.09497E-4

-1.09497E-4     2.00680E-3
```

```
THE PRODUCT OF THE NONHOMOGENEOUS VECTOR AND C INVERSE IS

 94.542

 6.516
THE COLUMN VECTOR OF THE COEFFICIENTS IS

 4.31636E-3

 2.72431E-3


TIME:  1.20 SECS.
```

### Exercise 11-6

Repeat the procedure in Program 11-5 so as to generate the entire matrix of coefficients from the information given in Exercise 11-5.

Solution 11-6. A second run with the new nonhomogeneous vector {0.401, 0.568, 0.209, 0.740, 0.042} leads to the solution vector at 525 nm $\{3.85 \times 10^{-4}, 4.29 \times 10^{-2}\}$. The coefficient matrix of absorbancies for the system is

$$\begin{bmatrix} 4.32 \times 10^{-3} & 2.72 \times 10^{-3} \\ 3.85 \times 10^{-4} & 4.29 \times 10^{-2} \end{bmatrix}$$

This result is in good agreement with the coefficient matrix given in Exercise 11-5 and obtained by single-variable linear regression. Multiple regression is a useful addition to the life scientist's computing repertory, particularly in situations in which it is impossible or undesirable to study the influence of one independent variable in the absence of all the other independent variables in a function of $n$ variables.

## Glossary

*Element.* Member of a rectangular array called a matrix. Elements are often numbers, but they are not restricted to numbers.

*Linear Regression.* Procedure for obtaining the best straight line through a set of data pairs, commonly a least squares procedure.

*Matrix.* Rectangular array of elements.

*Multiple Regression.* Procedure for obtaining the best set of equations for a function of more than one variable.

*Principal Diagonal.* Diagonal from the upper left corner of a matrix to the lower right corner.

*Regression Line.* Curve resulting from a linear regression technique for curve fitting, commonly a least squares method.
*System Subroutine.* Subroutione, generally supplied by the computer vendor, that is part of the system software, e.g., EXP and MATPRINT.
*Unit Matrix.* Matrix with 1 on the principal diagonal and 0 as all other elements.

## Problems

*1.* Solve the following simultaneous equations by Gaussian substitution

$$x_1 + x_2 = 5$$

$$x_1 - x_2 = 1$$

*2.* Write out a pair of nonhomogeneous independent simultaneous equations in two unknowns. Write out a pair of homogeneous simultaneous equations in two unknowns. Write out a pair of nonhomogeneous dependent simultaneous equations in two unknowns.

*3.* Obtain $A + B$, $A - B$, and $A \times B$ where

$$A = \begin{bmatrix} 1 & 2 \\ 3 & 4 \end{bmatrix} \quad B = \begin{bmatrix} 5 & 6 \\ 7 & 8 \end{bmatrix}$$

*4.* Write out the complete form of $C$ where

$$C = A \times B$$

and

$$A = \begin{bmatrix} a_{11} & a_{12} & a_{13} \\ a_{21} & a_{22} & a_{23} \\ a_{31} & a_{32} & a_{33} \end{bmatrix} \quad B = \begin{bmatrix} b_{11} & b_{12} & b_{13} \\ b_{21} & b_{22} & b_{23} \\ b_{31} & b_{32} & b_{33} \end{bmatrix}$$

*5.* Solve the following set of simultaneous equations by Cramer's rule

$$a + 2b + 3c = 0$$

$$4a + 5b + 6c = 1$$

$$7b + 8c = 0$$

*6.* Invert the matrix

$$\begin{bmatrix} 1 & 1 \\ 1 & -1 \end{bmatrix}$$

and demonstrate that you have obtained the correct matrix for $A^{-1}$.

*7.* Solve the following set of simultaneous equations by computer

$$5u + v + 3w = 16.0$$

$$u + 4v + w + x = 11.0$$

$$-u + 2v + 6w - 2x = 23.0$$

$$u - v + w + 4x = -2.0$$

*8*. Suppose two metal complexes have absorbancies of 3.50 at 400 nm and 0.500 at 650 nm in the visible for one complex and 0.500 at 400 nm and 2.500 at 650 nm for the other. All absorbancies are in units of L $mol^{-1}$. A mixture has $A = 0.450$ at 400 nm and 0.550 at 650 nm. What is the concentration of each complex in the mixture? Give units. Assume $b = 1.00$.

*9*. Write out $X^{TR}XM$ and demonstrate that it is indeed Expression (11-45). Expression (11-45) is a vector. Explain.

*10*. Use the rules of matrix multiplication to multiply the matrix

$$X = \begin{bmatrix} 2 & 3 \\ 1 & 5 \\ 0 & 1 \end{bmatrix}$$

into its transpose, i.e., obtain $XX^{TR}$.

## Bibliography


J. R. Barrante, *Applied Mathematics for Physical Chemistry*, Prentice-Hall, Englewood Cliffs, N.J., 1974.

C. T. Kenner, *Instrumental and Separation Analysis*, Merrill, Columbus, Ohio, 1973.

A. P. Gifford, S. M. Rock, and D. J. Comaford, *Anal. Chem.* **21,** 1026 (1949).

C. E. Meloan and R. W. Kiser, *Problems and Experiments in Instrumental Analysis,* Merrill, Columbus, Ohio, 1963, p.7.

M. G. Natrella, *Experimental Statistics*, National Bureau of Standards Handbook 91, U.S. Government Printing Office, Washington, D.C., 20402, 1966.

S. Chatterjee and B. Price, *Regression Analysis by Example*, Wiley, New York, 1977.

# Answers to Problems

## Chapter 1

*1*. (a) 10, (b) 13, (c) 43, (d) 119
*2*. 0011 + 0110 = 1001
*4*. 256, 65536

## Chapter 2

*1*. (a) 28.17, (b) 6.61E24, (c) 1.93
*2*. (a) 1.2E-3, (b) 5.2E10, (c) 16.8
*3*. 40.31, 100.5, 1364, 50960 (or 50950)
*4*. $\Delta A/A = \Delta B/B + \Delta H/H$
*5*. $dA = 0.5HdB + 0.5BdH$
*6*. Range: 1.399–1.456, mean = 1.430, median = 1.432 or 1.433, no mode, quad mean = 1.430
*7*. Sum $x^2$ = 1.5E2
*8*. Conclusion: The approximation is very good.

## Chapter 3

*1*. Range = 3.8–8.1
*3*. Mean = 37.0, $s$ = 3.8; sample statistics
*4*. Under the hypothesis stated, 37.0 and 3.8 become estimates of the population parameters.
*5*. Mean = 103.5, $d$(average) = 0
*6*. Mean = 41.0 cc/100 cc, $s$ = 2.5 cc/100 cc

## Chapter 4

*1*. 1/8 or 0.125
*5*. $p(0) = 0.36$
$p(1) = 0.48$
$p(2) = 0.16$

6. $p$[(7 or 11) and T] = 0.111
   $p$[7 or (11 and T)] = 0.194
7. $p$(4 or $J$) = 0.154
   $p$(4 and $J$) = 0.00591
8. $p(f,f)$ = 0.0324
9. 20, 10
10. $p(b,b)$ = 0.155
11. $p$(2 of 3) = 0.096
    $p$(at least 2) = 0.104

## Chapter 5

2. $f(0) = 0.579$
   $f(1) = 0.347$
   $f(2) = 6.94E-2$
   $f(3) = 4.63E-3$
3. $p = 0.01234$
4. 0.03125, 0.15625, 0.3125, 0.3125, 0.15625, 0.03125
6. 0.729, 0.243, 0.027, 0.001; most prob. failures: 3
9. $p$(3 survivals) = 0.008 due to chance alone. State the treatment is effective with 99% confidence.
10. Most probable survival number: 3
    $p(3) = 0.412$, $p$(at least 3) = 0.652
    The evidence is inconclusive.
11. The curve has a maximum at $p$(2 of 3) = 0.44 when $p$(first trial) = 2/3

## Chapter 6

3. $p = 0.175$
4. 1, 1.5, 1.71667, 1.71828, 1.71828
   The limit is $(e - 1)$.
7. Half time = $1.0E5$ $y$
8. $3.91E6$ atoms, $1.14E-12\%$, cpm(mean) = 832, $\langle s \rangle = 29$

## Chapter 7

1. (a) 0.4514, (b) 0.5289, (c) 0.4106
2. $z = -0.933$
3. $z = -1.50$
   $A = 0.8664$
   CL = 87%
   $p$(error) = 0.13 or 13%
4. Type I, Type II
5. 95% yes, 99% no

6. The statement "abnormal cholesterol" may be made at 87% confidence, but no higher.
7. 0.05 or 5% of the total

## Chapter 8

1. Significant at >99%
2. Chi squared = 2.17
   Linear interpolation indicates that the difference is significant at about the 85% level, but not higher.
3. $z = 3.5$
   The difference is significant beyond the 99% level for both 1 and 2.
4. $z = 1.56$
   The difference is not significant at either level.
5. $z = 1.93$
   One tail: significant at 97.3%
   Two tails: signficant at 94.6%
6. Neither is significant.
7. Mean = 270.86 kcal/mol, 95% CL = 0.75, 99% CL = 1.03 kcal/mol
8. No signficant difference
9. Mean = 72.8, $s = 4.15$, 95% CL = 11.5
   $p(>80) = 0.042$, $p(<60) = 0.001$

## Chapter 9

1. Selected value = 17.9, mean = 17.88
2. det = −24
3. det = $aaa - acc - abb + bbc + bcc - abc$
4. $x = 29/9$, $y = 14/9$
5. Solution set = (1, −3, 7)
6. 2.27 millions/year. There were no sales when production started.
7. $r = 0.998$ rounded to 1.0
8. $A = 0.118E5*C$, $s = 0.015$, $C = 2.60E{-5}\ M$
9. $r = 0.995$, but graphical treatment shows method 1 superior

## Chapter 10

1. $P = 1.03E5$ torr mL
2. $k = 1.30E{-2}$ reciprocal days
3. $k = 1.7E{-2}$ reciprocal seconds
   $= 4.6E{-2}$ reciprocal hours
   $= 1.2E{-4}$ reciprocal years
4. $k = 0.081$ reciprocal hours, half time = 8.6 hours
5. $k = 8.33$ reciprocal seconds

7. $B = 0.261$, $C = 7.64E-4$
10. $S = 645 + 1.09T + 8.00E-3*T*T$ where $T$ is in degrees Centigrade

# Chapter 11

1. Solution set = (3, 2)
3. $$A + B = \begin{bmatrix} 6 & 8 \\ 10 & 12 \end{bmatrix}$$
$$A - B = \begin{bmatrix} -4 & -4 \\ -4 & -4 \end{bmatrix}$$
$$A * B = \begin{bmatrix} 19 & 22 \\ 43 & 50 \end{bmatrix}$$
5. $a = 5/18$, $b = 4/9$, $c = -7/8$
6. $$\begin{bmatrix} 0.5 & 0.5 \\ 0.5 & -0.5 \end{bmatrix}$$
7. Solution set = (1, 2, 3, −1)
8. $c(1) = 0.100$ mol/L, $c(2) = 0.200$ mol/L
10. $$\begin{bmatrix} 2 & 3 \\ 1 & 5 \\ 0 & 1 \end{bmatrix} \begin{bmatrix} 2 & 1 & 0 \\ 3 & 5 & 1 \end{bmatrix} = \begin{bmatrix} 13 & 17 & 3 \\ 17 & 26 & 5 \\ 3 & 5 & 1 \end{bmatrix}$$
$$X' X = \begin{bmatrix} 5 & 11 \\ 11 & 35 \end{bmatrix}$$
Generally speaking, matrix multiplication is not commutative.

# Subject Index

N



O



P